Value and Economy of Marine Resources

From the ***Seas and Oceans*** Set
coordinated by
André Mariotti and Jean-Charles Pomerol

Value and Economy of Marine Resources

Edited by

André Monaco
Patrick Prouzet

ISTE

WILEY

First published 2014 in Great Britain and the United States by ISTE Ltd and John Wiley & Sons, Inc.

ISTE Ltd
27-37 St George's Road
London SW19 4EU
UK

www.iste.co.uk

John Wiley & Sons, Inc.
111 River Street
Hoboken, NJ 07030
USA

www.wiley.com

Library of Congress Control Number: 2014953027

British Library Cataloguing-in-Publication Data
A CIP record for this book is available from the British Library
ISBN 978-1-84821-706-5

Contents

Foreword

We have been asked by ISTE to stimulate work in the area of the environment. Therefore, we are proud to present the "Seas and Oceans" set of books, edited by André Monaco and Patrick Prouzet.

Both the content and the organization of this collection have largely been inspired by the reflection, initiatives and prospective works of a wide variety of national, European and international organizations in the field of the environment.

The "oceanographic" community, in France and internationally – which is recognized for the academic quality of the work it produces, and is determined that its research should be founded on a solid effort in the area of training and knowledge dissemination – was quick to respond to our call, and now offers this set of books, compiled under the skilled supervision of the two editing authors.

Within this community, there is a consensus about the need to promote an interdisciplinary "science of systems" – specifically in reference to the Earth's own "system" – in an all-encompassing approach, with the aim of providing answers about the planet's state, the way it works and the threats it faces, before going on to construct scenarios and lay down the elementary foundations needed for long-term, sustainable environment management, and for societies to adapt as required. This approach facilitates the shift of attention from this fundamental science of systems (based on the analysis of the processes at play, and the way in which they interact at all levels and between all the constituent parts making up the global system) to a "public" type of science, which is finalizable and participative, open to decision-makers, managers and all those who are interested in the future of our planet.

In this community, terms such as "vulnerability", "adaptation" and "sustainability" are commonly employed. We speak of various concepts, approaches or technologies, such as the value of ecosystems, heritage, "green" technologies, "blue" chemistry and renewable energies. Another foray into the field of civilian science lies in the adaptation of research to scales which are compatible with the societal, economic and legal issues, from global to regional to local.

All these aspects contribute to an in-depth understanding of the concept of an ecosystemic approach, the aim of which is the sustainable usage of natural resources, without affecting the quality, the structure or the function of the ecosystems involved. This concept is akin to the "socio-ecosystem approach" as defined by the Millennium Assessment (http://millenniumassessment.org).

In this context, where the complexity of natural systems is compounded with the complexity of societies, it has been difficult (if only because of how specialized the experts are in fairly reduced fields) to take into account the whole of the terrestrial system. Hence, in this editorial domain, the works in the "Seas and Oceans" set are limited to fluid envelopes and their interfaces. In that context, "sea" must be understood in the generic sense, as a general definition of bodies of salt water, as an environment. This includes epicontinental seas, semi-enclosed seas, enclosed seas, or coastal lakes, all of which are home to significant biodiversity and are highly susceptible to environmental impacts. "Ocean", on the other hand, denotes the environmental system, which has a crucial impact on the physical and biological operation of the terrestrial system – particularly in terms of climate regulation, but also in terms of the enormous reservoir of resources they constitute, covering 71% of the planet's surface, with a volume of 1,370 million km^3 of water.

This set of books covers all of these areas, examined from various aspects by specialists in the field: biological, physical or chemical function, biodiversity, vulnerability to climatic impacts, various uses, etc. The systemic approach and the emphasis placed on the available resources will guide readers to aspects of value-creation, governance and public policy. The long-term observation techniques used, new techniques and modeling are also taken into account; they are indispensable tools for the understanding of the dynamics and the integral functioning of the systems.

Finally, treatises will be included which are devoted to methodological or technical aspects.

The project thus conceived has been well received by numerous scientists renowned for their expertise. They belong to a wide variety of French national and international organizations, focusing on the environment.

These experts deserve our heartfelt thanks for committing to this effort in terms of putting their knowledge across and making it accessible, thus providing current students with the fundaments of knowledge which will help open the door to the broad range of careers that the area of the environment holds. These books are also addressed to a wider audience, including local or national governors, players in the decision-making authorities, or indeed "ordinary" citizens looking to be informed by the most authoritative sources.

Our warmest thanks go to André Monaco and Patrick Prouzet for their devotion and perseverance in service of the success of this enterprise.

Finally, we must thank the CNRS and Ifremer for the interest they have shown in this collection and for their financial aid, and we are very grateful to the numerous universities and other organizations which, through their researchers and engineers, have made the results of their reflections and activities available to this instructional corpus.

André MARIOTTI
Professor Emeritus at University Pierre and Marie Curie
Honorary Member of the Institut Universitaire de France
France

Jean-Charles POMEROL
Professor Emeritus at University Pierre and Marie Curie
France

1

The Services Provided by Marine Ecosystems: Economic Assessments and Their Usages

1.1. Marine ecosystem services

1.1.1. *Ecosystem services*

According to the *Global Footprint Network*, humanity is, metaphorically speaking, contracting a considerable "debt" to the environment because our consumption is not sustainable in the long term. This results in an erosion of the natural capital on which we depend to feed, warm, hydrate and house ourselves, and to engage in leisure activities. To render this understandable, this organization has established, based on the use of the ecological footprint indicator, that humanity has consumed a year's worth of natural capital by a point in the year situated around the middle of August. Between this date and the end of the year, humanity lives on with a debt. To make this idea of ecological debt even more coherent, a large number of scientists and also stakeholders in civil society have turned to the notion of "ecosystem service".

Chapter written by Harold LEVREL, Pedro CABRAL, Océane MARCONE and Rémi MONGRUEL.

Ehrlich and Mooney [EHR 83] seem to have been the first to mention the notion of ecosystem service explicitly in an article entitled "Extinction, substitution and ecosystem services". But it would be necessary to wait 14 more years to see this concept benefit from intense publicity through two widely-circulated publications: "Nature's services: Societal dependence on natural ecosystems" [DAI 97] and "The value of the world's ecosystem services and natural capital" [COS 97].

"The services that ecosystems offer are the benefits that people take from the ecosystems" (*Millennium Ecosystem Assessment*, [MEA 05, p. 9]). Ecosystem services contribute to human well-being from access to the essential goods that they provide (food, drinking water, etc.), the security that they offer (security against hazardous events, mitigation of the effects of climate warming, etc.) or simply the pleasure that they provide (observation of natural countryside, recreational activities in the open air, etc.).

The *Millennium Ecosystem Assessment* [MEA 05] carried out between 2001 and 2005 under the auspices of the UN and involving 1,360 scientists aimed to describe these ecosystem services precisely. The MEA is distinguished into the following four categories of ecosystem services:

– provisioning services equate schematically to the natural resources that are used through a process of extraction for mankind's direct consumption;

– regulating services, which represent ecological functions enabling the productivity and resilience of ecosystems to be guaranteed;

– cultural services, which are both recreational (activities in the open air) and subjective in nature (spirituality, identity, etc.);

– supporting services, in conjunction with ecological processes, enable the renewal of life on Earth.

Through these categories, a new approach of the ecological and economic dynamics is available to us. The approach of using the idea of ecosystem services effectively allows us to put forward an unprecedented discourse on the conservation of biodiversity by underlining the trade-offs that are necessary to make between the different types of services furnished by biodiversity and means that the process of economic development and biodiversity conservation objectives are no longer systemically opposed. The notion of ecosystem service also provides a common semantic and theoretical base for different disciplines to work on the problem of interaction between the question of conservation and the question of development, and also a unit for assessment that allows interaction with decision-making bodies [DAI 08, RUF 09].

The MEA focused on ecosystem services but also on the pressures that are exerted on them. In effect, it underlines which human activities today cause the greatest threats to these services through their consumption of space, their exploitation of resources, the emission of greenhouse gas or the introduction of invasive species. Through their activities, people destroy a significant quantity of ecosystem services and thus, finally, their collective well-being (Figure 1.1).

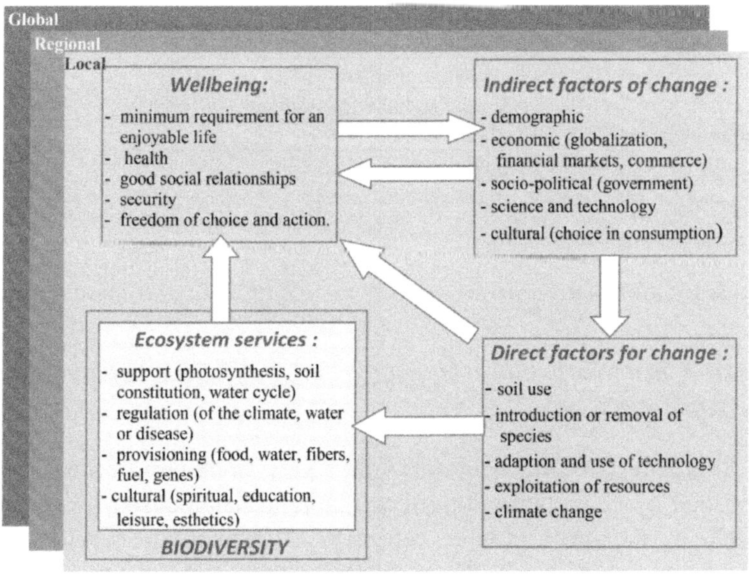

Figure 1.1. *The Millennium Ecosystem Assessment*

Category of services	Services	Evolutions
Services for provisioning or supply	Agriculture	+
	Farming	+
	Fishing	–
	Aquaculture	+
	Wild food materials	–
	Wood for construction	±
	Cotton, hemp, silk	±
	Firewood	–
	Genetic resources	–
	Biochemical products, natural medicines, pharmaceutical products	–
	Freshwater	–
Regulating services	Regulation of air quality	–
	Regulation of global climate	+
	Regulation of regional and local climate	–
	Regulation of the water cycle	±
	Regulation of erosion	–
	Purification of water and treatment of waste	–
	Regulation of diseases	±
	Regulation of parasites	–
	Pollinization	–
	Regulation of natural risks	–
Cultural services	Spiritual and religious values	–
	Esthetic values	–
	Recreation and ecotourism	±

"+" stands for an increase, "–" stands for a reduction and "±" stands for an increase in certain regions of the world and a drop in others. The support services are not mentioned here, as it is difficult to evaluate their evolution.

Table 1.1. *Evolution of ecosystem services between 1950 and 2005 [MAE 05, p. 46]*

1.1.2. *A historic balance leading to an inefficient exploitation of ecosystem services*

At the present time, man only uses a very small part of the services furnished by ecosystems. In effect, mankind has focused for the past 10,000 years on a single category of ecosystem services: the "provisioning services". The regulating and cultural services have always been neglected in favor of provisioning services. Thus, Table 1.1, which succinctly summarizes the evolution of the main ecosystem services over the course of the second half of the 20th Century, allows us to highlight that regulating services and cultural services have, for the most part, been neglected in comparison with provisioning services. Even if this tendency is easily justified by the need to keep up with the dramatic demographic increase that the world has known in the course of this period, the productivist model that is at its origin is today being pushed to its limits, and one of the main indicators that allows us to note this is the erosion of regulating services, which are also, indirectly, absolutely necessary for the human species.

This focus on ecosystem provisioning services has led to great inefficiency in the exploitation of biodiversity. Thus, most of the time, the strategy for biodiversity exploitation is based on the desire to maximize the production of particular provisioning services and neglect to take into account other categories of ecosystem service.

The marine ecosystem is a perfect example of this. A marine ecosystem will, in effect, be most often used to provide a single ecosystem service – fish for food – although this ecosystem is fundamental for a large number of ecosystem services for man – climate regulation, providing a habitat for species, a place for recreational activities, molecules or genes for the development of medication, etc.

The inefficient use of biodiversity and the services that it provides is one of the first factors that explains why our mode of development is not sustainable. The MEAs conclusions are indisputable. They emphasize that 60% of ecosystem services have decreased during the last 50 years. Among them, the renewal of fishing stocks and the production of freshwater seem to be most threatened. This degradation has been more significant over the course of the last 50 years than over the entire course of the rest of human history, and it will be even more significant in the 50 years to come. The ecosystem services that are disappearing are those of a collective or public nature and those that are not sold on the markets (recycling of waste, reproduction habitats for animals or countryside for mankind, etc.). Conversely, those that have been developed over the last 50 years are services of a private nature that can be sold on the markets and which today form the basis of the forestry, agriculture and aquaculture sectors.

This is why our systems for exploiting nature need to undergo radical change and take into account the ensemble of ecosystem services and most particularly communal or public cultural and regulating ecosystem services.

1.1.3. *Marine ecosystem services*

Ecosystem services with coastal habitats as their origin are so numerous that they could account for 43% of the ensemble of services furnished by the biosphere, although coastal ecosystems only represent 6.3% of the globe's surface [COS 97].

For France, the relative importance of ecosystem services associated with the sea and the seashore for the national economy can be evaluated as an initial approximation via the source revenue generated by these areas. This initial approximation is particularly simple for provisioning and recreational services. Thus, the sea product industry generated an added value of 2.363 billion euros in 2007, whereas the coastal tourism industry generated 9.220 billion euros for a total added value of activities depending on marine ecosystems state 27.6 billion euros (*Données économiques maritimes 2009*). The maritime economy is moreover the source of 486,000 jobs, including 242,558 linked to the tourism sector. An important part of this wealth is linked to the provision of services by marine biodiversity.

It is in effect the more or less direct origin of numerous ecosystem services such as:

– bioturbation, primary production or the water cycle (supporting services);

– the renewal of fisheries, the production of aragose (derived from algae) or renewable energy (provisioning services);

– the control of erosion and silting, the recycling of waste and the control of pollution (regulating services);

– visual tourism, recreational fishing or simply bathing (cultural services).

By relying on a review of the literature, it is possible to identify 74 services directly linked to marine and coastal biodiversity, including seven for support services, 20 for provisioning services, 27 for regulating services and 18 for cultural services ([COS 97, DUA 00, HOL 99, JAC 01, KAI 11, KRE 05, MEA 05 (Chapters 18 and 19), RON 07, SOL 04, WOR 06]; Tables 1.2–1.5).

Support services	Source of service's production (structure or function)
Bioturbation	Benthic invertebrate species biodiversity and fish that carry out activities in the substrate (spawning, searching for food, sheltering)
Primary productivity	Genetic and species biodiversity through a complementarity effect of redundance and selection
Secondary productivity	Genetic and species biodiversity through a complementarity effect of redundance and selection
Cycle of nutrients and mineralization	Gas fixing and decomposition of organic matter by species biodiversity, fundamental to the nitrogen production necessary for primary production
Water cycle	Oceans as an essential base of the water cycle
Creation of habitats for animals and vegetables (formation of soils)	Biodiversity of soil invertebrates, microorganisms in the soil, nitrogen fixing plants, plants and animals that produce organic waste
Oxygen and carbon cycle	Oceans as an essential base of the oxygen and carbon cycles

Table 1.2. *Marine ecosystem support services and their sources (n = 7)*

Provisioning services	The service's source of production (structure or function)
Renewable energy	Marsh, swell, current, oil from microalgae
Fish for food	Genetic and species biodiversity helps limit the risks of extinction for fisheries as well as the variability of takings and facilitates the renewal of fisheries in crisis
Crustaceans for food	Abundance and diversity of crustaceans
Molluscs for food	Abundance and diversity of mollusks
Algae and derivatives (agarose) for food	Abundance and diversity of algae
Materials for construction	Abundance and diversity of primary materials necessary for construction (marine sediments, sand and stones, nodules)
Materials for clothing	Abundance and diversity of primary materials necessary for clothing (skins, viscera, etc.)
Non-renewable energy	Abundance and diversity of petrol and gas deposits
Spat (for farmed shellfish)	Spat biodiversity
Fish meal for animal food	Fish biodiversity
Fish oil for animal food	Fish biodiversity
Fertilizer	Biodiversity enabling organic fertilizers to be produced (algae, kelp, fish bone, etc.)
Molecules for pharmaceutical products	Molecular diversity of renewable and non-renewable resources in coastal and marine zones
Chemical models	Biodiversity possessing the necessary characteristics to provide organisms with tests
Genetic resources	Marine and coastal genetic biodiversity
Materials for artistic productions (pearls, mother of pearl, etc.)	Biodiversity that gives rise to the production of materials useful for artistic productions
Support for the transport of merchandise and passengers	Oceans and seas as supports for marine routes
Organisms with tests	Biodiversity of organisms possessing the necessary characteristics to provide organisms with tests
Molecules for industrial and cosmetic products (glues and creams)	Molecular diversity of renewable and non-renewable resources from marine and coastal zones
Freshwater storage (estuaries)	Estuaries in good ecological state

Table 1.3. *Marine ecosystem provisioning services and their sources (n = 20)*

Regulating services	The service's source of production (structure or function)
Dynamic of soil fertility	Biodiversity of soil invertebrates, soil microorganisms, nitrogen-fixing plants, plants and animals that produce organic waste
Control of the phytoplankton dynamic	Diversity and abundance of zooplankton
Control of the zooplankton dynamic	Diversity and abundance of zooplanktivores
Control of the dynamic of fish populations	Diversity and abundance of piscivores
Maintenance of hydrological equilibrium	Water cycle assured by the oceans and coastal habitats
Spawning grounds for species	Biodiversity of seagrass, wetlands, salt marshes and oyster beds
Refuge zone for species	Diversity of marine and coastal habitats
Resilience in the face of natural or human disturbances	Genetic and species biodiversity through the effect of complementarity, redundance and selection
Control of pathogens and harmful materials	Species biodiversity via the role of populations of predators and filtering organisms
Regulation of herbivores	Species biodiversity via the role of populations of predators
Mitigation of the eutrophization effects	Species biodiversity via the role of filtering organisms
Control of pollution and detoxification	Species biodiversity via the role of filtering organisms
Transfer of energy from the substrate to higher trophic levels	Species biodiversity that makes up the food chain
Control of waves and energy currents	Diversity of marine and coastal habitats that act as buffer zones (seagrass, dunes, etc.)
Control of erosion and silting	Seagrass biodiversity
Mitigation of the effects of rising sea level and floods	Species biodiversity in vegetation (mangroves)
Protection against ultraviolet	Oceans and seas have an important role in the biogeochemical cycles and shelter microorganisms useful for protection against UV

Air purification	Oceans and seas have an important role in the oxygen and carbon cycles
Regulation of the global climate	Oceans stabilize the quantity of CO_2 in the atmosphere and regulate the temperature of the global atmosphere
Regulation of local climates	Oceans stabilize the quantity of CO_2 in the atmosphere and regulate the temperature of the local atmosphere
Retention of soils	Root species biodiversity
Control of water turbidity	Species biodiversity via the role of filtering organisms
Regulation of water quality	Species biodiversity via the role of filtering organisms
Control of human diseases	Microbiological diversity
Transport of species	Currents and tides
Recycling of waste	Species biodiversity of soil invertebrates and aquatic microorganisms
Regulation of salinity	The salinity levels of coastal zones are dependant on fluxes of freshwater from land
Carbon stocking	Biodiversity of phytoplankton, macroalgae and seagrass

Table 1.4. *The marine ecosystem regulating services and their sources (n = 27)*

Cultural services	Source of the service (structure or function)
Feelings of well-being	Marine and coastal ecosystems
Support for "traditional" jobs for coastal populations	Abundance of resources on which local communities depend
Cultural identity of coastal populations	Natural coastal countryside in connection with traditional practices
Views (coastal countryside)	Countryside biodiversity
Ecotourism	Countryside biodiversity
Visual tourism (whales, dolphins)	Animal biodiversity
Bathing	Biodiversity of filtering species
Walking	Countryside biodiversity

Recreational fishing	Biodiversity in species valued for recreational fishing
Pleasure sailing	Seas and oceans devoid of very large floating objects (invasive species, macrowaste, other boats) and without excessive eutrophization
Deep-sea diving	Marine biodiversity
Surfing and windsurfing	Waves and wind
A source of inspiration	Countryside biodiversity
Support for religious beliefs	Sacred natural places and objects
Preservation of marine and coastal biodiversity for ethical reasons	Marine and coastal biodiversity
Source of knowledge	Marine and coastal biodiversity
Scientific usage (a marine model for basic research)	Marine and coastal biodiversity
School excursions	Marine and coastal biodiversity
Monitoring of global changes affecting the natural environment	Monitoring of phenological characteristics and species distribution

Table 1.5. *Marine cultural ecosystem services and their sources (n = 18)*

1.2. The monetary evaluation of ecosystem services

1.2.1. *The factors that motivate demands for monetary evaluation*

1.2.1.1. *The demands for monetary evaluation of ecosystem services in an institutional framework*

From the 1990s, the evaluation of ecosystem services has been recommended as a tool to aid decision making on the question of biodiversity, and this applies within a variety of governing bodies.

Different international organizations therefore promote the use of economic evaluations of ecosystem services. In its IV/10 decision, the Conference of the Parties (COPs) at the Convention on Biological Diversity (CBD) considers that "the economic evaluation of biological diversity and biological resources constitute an important tool for well-targeted and well-distributed economic incentives" [CON 98]. In its principle 4, the COPs decision VII/11 calls again for "the incorporation of the ecosystem and social aspects of goods and services resulting from ecosystems in decisions relating to national compatibility, politics, planning, education and the management of resources" [CON 04, p. 217].

In 2007, a report on biodiversity from the Parliamentary Office for the assessment of scientific decisions [LAF 07] underlined that "the sustainable development of biodiversity is a necessity and an opportunity. Two axes are profiled in this domain: the remuneration of services provided by ecosystems and the exploration of a reservoir of goods that could be a key tool for the fourth industrial revolution …: it is necessary to evaluate ecosystem services monetarily and to provide economic sanctions for their destruction for private ends".

More recently, the OECD (Organization for Economic Co-operation and Development) has recalled that "appropriate economic evaluations of biodiversity and its loss will result in better, more efficient decisions, and can prevent inappropriate compromises" [OEC 12, p. 191].

From the point of view of many political decision makers, the monetary evaluation of biodiversity seems to have become a tool that could better help protection of biodiversity, and this point of view is not unique to economists. Conservation biologists, and more broadly environmental non-governmental organizations, defend this approach to show that biodiversity "is worth" something. Public administrative bodies in charge of environmental policies highlight this approach to allow using new financial methods or for optimizing their projects for public development. Economists seek to place monetary units on ecosystem services to promote work on the question of biodiversity and to put in place tools for market regulation. For most institutional bodies, it ultimately involves the most used standard of measurement in a society with a market economy and we cannot avoid it when discussing the conservation of biodiversity.

To resume, the key argument to justify granting a monetary value to biodiversity is that if an ecosystem service has no monetary value, it will at best remain unused and at worst will be wasted. And it is evident that public representations are strongly influenced by the monetary standard, which is the most used indicator for transactions in our market society. Thus, the monetary evaluation of ecological phenomena would offer a strong tool for argument in societies with market economies.

Then, from a very pragmatic point of view, the question of the monetization of biodiversity and the services that it provides seem essential in many cases.

First, for insurance, because the monetary value enables the economic risks associated with the destruction of biodiversity to be taken into account, monetization should in particular enable us to evaluate the indemnity linked to external factors (e.g. pollution by hydrocarbons or increased risk of flooding), but also, possibly, premiums for good practice.

Then, for fiscal policies, monetization seems necessary for putting in place systems for taxation and subsidies that should ideally reflect the societal costs and benefits associated with the evolution of ecosystem services, with a view for creating the necessary incentives and leading to changes in behavior.

Finally, for the choice between public or private investments, including the question of biodiversity in investment, decisions require the ability to carry out cost-benefit evaluations for the different projects in order to be able to compare and prioritize them.

For the "market" finally, since biodiversity has a economic value, it can give rise to commercial exchanges, can be valued and can inspire investment in its preservation and restoration.

1.2.1.2. *The regulatory effect of the monetary evaluation of ecosystem services*

There are two opposing theories concerning the regulatory effect of establishing a price for biodiversity. According to Timothy Swanson's analysis [SWA 94], a high price would be an incitement to conserve a natural renewable resource since it is imperative not to "kill the goose that lays the golden eggs". On the other hand, according to Clark's analysis [CLA 73], the placing of a higher price on renewable natural resources would trigger a rapid and unsustainable use of the latter. In particular, practices on the luxury market can be observed, which do not necessarily respond to conventional economic rules. So, the increasing rarity of a consumed species (for food, collection, etc.) will cause the costs of sourcing it, and therefore the market cost, to increase this without causing a drop in demand. There is an anthropogenic "allee effect" with which we can underline the value that mankind attributes to rare species and which accounts for their exploitation down to the last individual Franck Courchamp *et al.* [COU 06]. Gault *et al.* [GAU 08] emphasize, moreover, that this behavior is based above all on the perception of rarity and not on real biophysical rarity.

In this respect, the two theories are exactly the same. In effect, high or low prices can have inverse effects depending on the contexts in which they occur. The important element in this context is access regulation. So, in cases where a natural renewable resource has a strong monetary value, the consumers who benefit from this exclusive access will tend to seek a management method that assures the effective renewal of this source of revenue. On the other hand, if access is not secure, or indeed free, then it is rational to use it down to the last unit of resource so long as there is a solvent demand for this resource.

1.2.2. *Monetary evaluation methods and their limits*

1.2.2.1. *The values of ecosystem services*

To respond to institutional demands concerning the monetary value of biodiversity, the *The Economics of Ecosystem and Biodiversity* (TEEB) was launched in 2007 (www.teebweb.org) through the initiative of the G8 and five developing countries. The goal of this work was to achieve a better understanding of the economic benefits resulting from biodiversity. The TEEB "promotes the integration of economic values for biodiversity and the services provided by ecosystems in the decision-making process" [TEE 10, p. 27].

But the economic evaluation of ecosystem services began much earlier than the TEEB, which was instead an opportunity to take stock of what already existed. Thus, Laurans *et al.* find 5,028 references issued from 1,419 sources corresponding to the following key words: "evaluation" and "ecosystem services", "natural capital", "environment" and "evaluation", "biodiversity" and "evaluation" and "total economic value"[1] [LAU 13, p. 210].

For marine ecosystems, a quick search on ScienceDirect using the key words "ecosystem services", "evaluation" and "marine" returns around 1,026 references[2].

The creation of economic evaluations of marine ecosystems is not only found in the academic literature. So, governments and governmental agencies, non-governmental organizations and *think tanks* also produce economic evaluations. In order to collect these evaluations, the database *Marine Ecosystem Services Partnership* (www.marineecosystemservices.org/explore) aims to collect existing economic evaluations on marine ecosystems, including the gray literature.

The idea of putting a price on ecosystem services to take account of the consequences of environmental degradation on social well-being is based on a utilitarian principle [BON 07]. This approach depends on the notion of sustainability, which enables us to derive an economic value reflecting individuals' attachment to the different goods and services to which they have access. Thus, according to a economic value, it implies that environmental breakdown has an impact on the utility functions of individuals. It is this impact that is important to measure in monetary terms in order to offer adapted price signals.

To do this, a classification of the different forms of value that biodiversity can take is required [BON 07]:

– the use value equates to a direct and current use of the asset;

---

1 Results on 31/01/2012, the three databases ISI (access via *Web of Science*) and Elsevier's Scopus have been consulted, key words in English.

2 Results on 13/12/2013, carried out on ScienceDirect Elsevier, key words in English.

– the non-use value equates to a direct, future use of the asset by the present generation (option value) or for future generations (legacy values);

– the existence value equates to the value of the asset for its own sake, regardless of usage.

The sum of these three types of value corresponds to the total economic value of biodiversity (Figure 1.2).

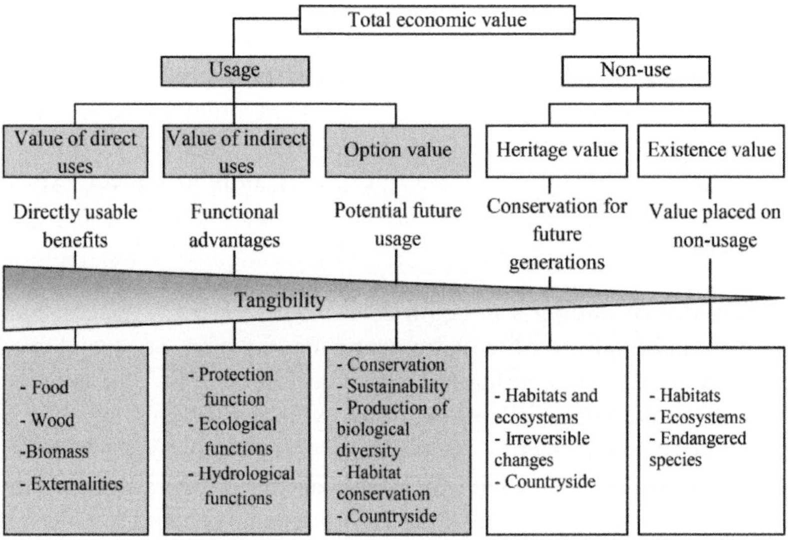

Figure 1.2. *The total economic value of biodiversity and ecosystem services (source: Centre d'analyse stratégique, 2009)*

It is evidently considered possible to evaluate monetarily the benefits provided by ecosystem services and the biodiversity that generates them – hence the desire to give them a monetary value. However, to carry out the monetization of ecosystem services is very much complex.

If provisioning services are associated with exploited resources for which there is a market price, the three other categories of services provided by the ecosystems are not, for much of the time, the object of any commercial transaction and for this reason cannot be measured in monetary terms using a market price. Also, different methods have been developed by environmental economists to try, in spite of everything, to attribute a monetary value more or less directly to these ecosystem services.

1.2.2.2. *Assessment methods for regulating services and cultural, recreational services*

Where regulating services are concerned, methods often rely on real market processes: by estimating the contribution of regulating services to the creation of products and services sold on the markets (for example the contribution of breeding habitats for the fish that make up fishing stock); by estimating the cost of restoration or damage prevention to regulating services important for society (for example the cost of putting in place a dyke to replace the protecting function furnished by destroyed coral ecosystems). It is a question, above all, of creating a link between natural habitats and the production of ecosystem services, directly or indirectly useful to mankind.

For recreational cultural services, environmental economists have adopted the reconstitution method for market price. The reconstitution of market price is based either on the travel costs method or on the hedonic pricing method. The travel costs method consists of evaluating the time that some individuals and groups are prepared to "spend" to benefit from the services offered by a particular ecosystem [APP 04]. The hedonic pricing method is based on the hypothesis that some goods – notably housing – have a value that does not depend only on the characteristics of the object itself, but also on the natural environment in which the latter is situated [TRA 08].

Although these assessment methods carried out using preferences revealed by the market are relatively robust, they are hampered by two limitations.

The first is the cost of developing them, since they require information to be collected on site and since they use a large quantity of very precise data and require the use of relatively sophisticated econometric models to be able to put forward a robust assessment.

The second limit of these methods is that they do not enable us to take account of values for the non-use of biodiversity, nor the option values, and only give a very vague suggestion of the values for indirect use. Indeed, the values of the ecosystem services associated with biodiversity seem to be linked largely to non-uses or indirect uses (Figure 1.2; [ENV 09, HEA 00, NAE 09]). This is why environmental economists have suggested creating hypothetical markets on which agents can express preferences that will not necessarily be linked to the uses of biodiversity, but to the value accorded to them for their own sake.

1.2.2.3. *The methods for assessing values for the non-use of biodiversity*

Today, the most used method for assessing the non-use values of biodiversity in monetary terms is the contingent valuation method (CVM) [BON 07]. The CVM is based on preferences concerning the conservation of biodiversity declared in the context of a hypothetical market. To make this hypothetical market more coherent, interviewees were asked how much they would be willing to receive in return for one or several units of particular components of biodiversity destroyed or, conversely, how much they would be willing to pay to improve the restoration of one or several units of these same components. Thus, through the means of statistical analyses enabling a large number of variables to be taken into account relative to the context in which these preferences are expressed, as well as the personal parameters of the respondent, it is possible to accord values to the biodiversity components concerned. The advantage is evidently that individuals can express preferences for the biodiversity components that do not arise from any direct use but simply to the fact that they attribute a value to biodiversity for its own sake.

These methods are however the source of significant controversies on the subject of their capacity to estimate non-use values [OCO 99, HEA 00]. Milanesi [MIL 11] has published an article that summarizes these limits. His article emphasizes a publication by Diamond and Hausman dating from 1994 [DIA 94]; these authors suggest testing the coherence of the CVM by evaluating its precision, credibility and reliability using experimental methods. If the precision of the method is debatable, since it leads to considerable variability in the estimations, the authors do not consider this point to be a prohibitive problem for this method, since it is possible to increase the size of the sample up to the point where the confidence level is good (even if that has a considerable cost). Similarly, if the CVM generates important biases that can limit its reliability, it is technically possible to control these biases using a strict protocol for the enquiry and statistical corrections *ex post*. The credibility of the CVM seems, however, to pose a fundamental problem for the authors, who believe that the individuals surveyed are not responding to the question put to them since they do not understand, in the majority of cases, a trade-off based on a logic of substitution between the environmental good considered and their revenue. The responses reflect (1) an attitude that relates to a public good and not really to the question [KAH 94], (2) a feeling of moral satisfaction [KAH 92], (3) a personal cost–benefit calculation concerning the project or the regulations in question [PAY 93], or indeed (4) a reaction that seeks, for example, to "punish the polluter" [MIL 11, p. 13].

1.2.2.4. *The monetization frontier*

The factors outlined above explain why it is today accepted that if the values of the direct and indirect uses of the numerous ecosystem services provided by biodiversity are approached in a reliable manner, where these data exist, the values of non-uses (cultural and spiritual services) or those associated with the elementary processes of the reproduction of living organisms (supporting services) are, however, difficult to estimate for technical (a less tangible and familiar value in the eyes of individuals) and ethical (can one give a value to the elements that make up life on Earth [NOR 98]?) reasons.

There is therefore a limit beyond which the monetization of biodiversity and the services that it provides makes little sense [OCO 99]. The opportunities for monetary valuation concerning the non-commercial environmental functions become weaker; the broader the scales on which these evaluations should be carried out, the larger the complexity of the phenomenon to be evaluated and the greater the diversity of the values that can be attached to them (notably the non-use values). This limit to monetization should enable us "to distinguish clearly on principle what arises from deliberation (from social and political debate) and what can be handled by economic analyses" [VAN 02, pp. 439–440].

It therefore seems fairly logical to discard the non-use values in the assessments and simply seek to provide information on the value of ecosystem services reflecting specific uses, benefiting specific economic actors, on a defined/restricted spatial and temporal scale [HEA 00]. The gain thus obtained from the point of view of the reliability of the assessments allows us, however, to abandon the goal of aggregating the values and obtain a total economic value for biodiversity and the ecosystem services that it provides, without which the latter will mechanically underestimate the real value of the ecosystem services associated with the conservation of biodiversity.

A multi-criteria approach to the value of biodiversity and ecosystem services seems from this point best adapted to provide a broad range of indicators for ecosystem services based on ecological, economic or social information, which makes sense to a wide variety of actors. This tool can be used in collective decision-making processes and does not reduce this process to a cost–benefit analysis, enabling an optimal choice to be made in which all forms of value would be reduced to a single monetary standard.

1.3. The monetary evaluation of ecosystem services: some results for marine ecosystems

As mentioned above, the program entitled *Marine Ecosystem Services Partnership* (www.marineecosystemservices.org/, 11 December, 2013) has the objective of structuring reflections on an international scale around the evaluation of marine ecosystem services. On 11 December, 2013, it had collected 1,940 evaluations of marine ecosystem services corresponding to 893 studies.

Our analysis is based on a sample of 114 studies (containing 355) issued from this database reveals a considerable amount [MAR 13]. One-third of the studies evaluate food provisioning services (56 assessments, being 15.78% of the total) and services linked to tourism, recreational activities as well as esthetic values (63 assessments, being 17.75%, of which the vast majority focus only on tourism). The control of erosion and the protection of the shore represent 10.99% of the values produced by these studies; 5.64% of the assessments focus on soil formation (2.82%) and genetic resources (2.82%). Other ecosystem services form the object of 11–27 economic evaluations (Figure 1.3).

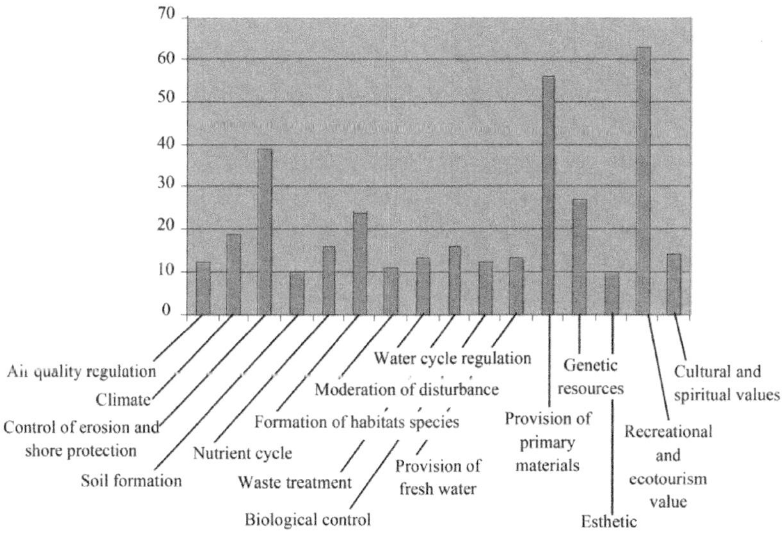

Figure 1.3. *Types of ecosystem services evaluated monetarily [MAR 13]*

The over-representation of provisioning services (food and construction materials) and cultural and recreational services can be explained by the existence of market prices or tangible indicators concerning the economic impact of these activities (visits, jobs, net benefits, etc.) making an economic evaluation easy to carry out. We can also underline the presence of a regulating service that stands out: the service of controlling erosion. This service is well represented since it is rather easy to measure the cost of coastal erosion, in monetary terms, in comparison with the lost real-estate value. Two other regulating services are frequently evaluated: refuge habitats and climate regulation. The former is necessary for the renewal of exploited fishing stocks. The second is in line with carbon storage, which can be translated into a monetary value due to the existence of a market for "carbon tons".

These results can be put into perspective using a literature review carried out by de Groot *et al.* [DE 12] in which the authors analyze 665 monetary values calculated for 22 ecosystem services furnished by 10 biomes. The results are presented in Table 1.6.

Within the 10 biomes presented, the marine and coastal biomes (coral reefs, coastal systems and wetlands) have higher total economic values. The highest values are those of regulating services. The regulation of erosion is the service that reaches the highest monetary value for coral reefs and coastal systems. In the case of coastal wetlands, it is the waste treatment service that has the greatest economic value. Conversely, the total economic value of the marine biome *stricto sensu* is the least significant of the 10 biomes. It should be noted that only 14 estimations (out of 665) were carried out for this biome, which concerns only five ecosystem services (for the other biomes, 10–17 ecosystem services are evaluated). This can be explained by the difficulties inherent in the monetization exercise on an area that is still little known.

To go further in the analysis, it is interesting to see how these values have evolved over time, by looking back at estimations conducted in 1997 by Costanza *et al.* and which were based on the same calculation methods described in the work of de Groot *et al.* ([MON 13]; Figure 1.4).

	Seas	Coral reefs	Coastal systems	Coastal wetlands	Inland waters	Fresh water (rivers, lakes...)	Tropical forests	Temperate forests	Wood	Grassland
Provisioning services	102	55724	2396	2998	1659	1914	1828	671	253	1305
1. Food	93	677	2384	1111	614	106	200	299	52	1192
2. Water				1217	405	1808	27	191		60
3. Primary materials,	8	21528	12	358	425		84	181	170	53
4. Genetic resources		33048		10			13			
5. Medical materials				301	114		1504		32	1
6. Ornamental materials		472								
Regulating services	65	171478	25847	171515	17364	187	2529	491	51	159
7. Regulation of water quality							12			
8. Climate regulation	65	1188	479	65	488		2044	152	7	40
9. Moderation of disturbance		16991		5351	2986		66			
10. Hydraulic regulation					5606		342			
11. Waste treatment		85		162125	3015	187	6	7		75
12. Control of erosion		153214	25368	3929	2607		15	5	13	44
13. Nutrient cycle				45	1713		3	93		
14. Pollinization							30		31	
15. Biological control					948		11	235		
Habitat services	5	16210	375	17138	2455	0	39	862	1277	1214
16. Plant nurseries		0	194	10648	1287		16		1273	
17. Genetic diversity	5	16210	180	6490	1168		23	862	3	1214
Cultural services	319	108837	300	2193	4203	2166	867	990	7	193
18. Esthetic value		11390			1292					167
19. Recreation	319	96302	256	2193	2211	2166	867	989	7	26
20. Inspiration		0			700					
21. Spiritual experiences			21							
22. Cognitive development		1145	22					1		
Total monetary value	**491**	**352249**	**28917**	**193845**	**25682**	**4267**	**5264**	**3013**	**1588**	**2871**

Table 1.6. *Summary of monetary values for each ecosystem service per biome (values in int.$/ha/year, 2007 price levels) [DE 12, p. 55]*

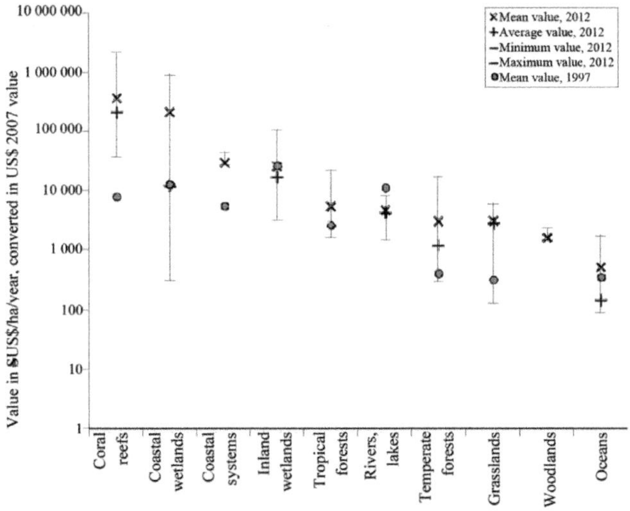

Figure 1.4. *Changes in monetary values observed between estimations by Costanza et al. in 1997 and those of de Groot et al. in 2012 (in US$ 2007)*

In Figure 1.4, it is shown that in 15 years, the changes in the values of ecosystem services provided by different types of natural habitat, representing the natural capital present on the Earth, have been very significant.

This variation seems to be the result of specific social and institutional factors. The article by de Groot *et al.* emphasizes, for instance, that the explosion in the value of ecosystem services associated with coral reefs (multiplied by a figure between 10 and 100 depending on the estimations) is mainly linked to increased recognition of the role played by this ecosystem in mitigating the phenomena of erosion. Thus, it is the growing perception of the risks associated with climate change and the consequences of the latter in terms of extreme events that has led to a much greater value to be attributed to all the habitats that can mitigate its effects. In the same way, the increase in the value of the climate regulating service seems to be a simple artifact of the creation of a carbon market (even if this value is dropping). The importance of this type of result is that it allows us to emphasize that the evolution of the value of ecosystem services is above all a question of perception and institutional changes, but not a consequence of the evolution of the ecosystem's dynamic as such.

A third and last level of analysis is the comparison of investment in the ecosystems and the return on the investment generated by the services whose creation they enable. This can be carried out by estimating ecological restoration costs and the benefits associated with ecosystem services thus produced. *In fine*, this should enable the calculation of internal rates of return (IRR) and benefit–cost ratios (BCR) for restoration projects for ecosystems.

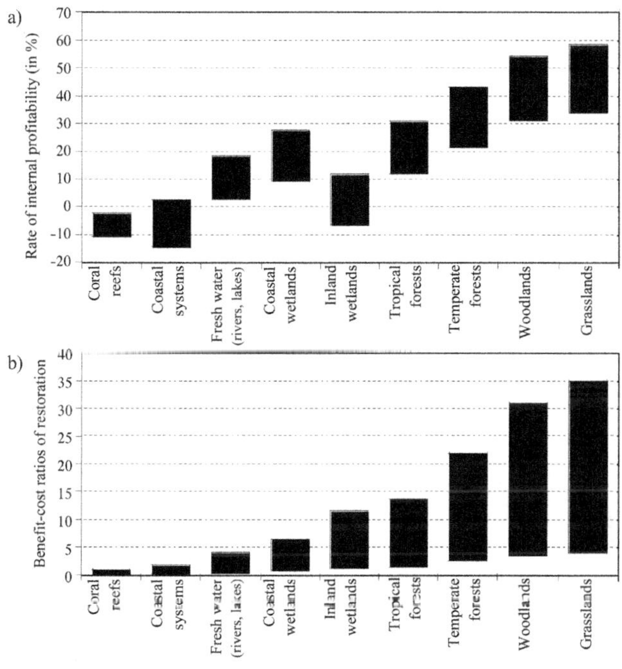

Figure 1.5. *a) Internal rate of return at 100% of the highest restoration costs reported and 30% of profits; the top of the bar represents the best case scenario (analysis conducted at 75% of the highest reported restoration costs and 75% of profits). b) Benefit–cost ratios of restoration (the bar represents the size of the values; the bottom of the bar, the worst case scenario [analysis conducted at 100% of the highest reported restoration costs, 30% of profits and a rate of social discount of 8%]; the top of the bar represents the best case scenario [analysis conducted at 75% of the highest reported restoration costs, 75% of profits and a rate of social discount of –2%]) in the nine main biomes based on 316 case studies over 20 years with management costs equivalent to 5% of capital [DE 13, p. 1290]*

Using 94 studies relating the restoration costs for ecosystems and 225 studies calculating the total economic values of the latter, de Groot *et al.* [DE 13] have calculated the IRR and BCR of restoration projects for nine biomes (Figure 1.5). The coral reefs and coastal systems have the weakest IRR and BCR, whereas their total economic values are most significant. This result is explained by the exceedingly high restoration costs.

Conversely, the grasslands have a relatively weak total economic value (Figure 1.4) but the restoration projects for these biomes offer significant returns on investment (Figure 1.5). The authors emphasize that "the restored ecosystems that offer the best return in absolute value [that is to say based on the current net value] are coastal wetlands and wetlands situated inland" [DE 13, p. 1289].

1.4. The effective use of the assessment of benefits associated with ecosystem services

1.4.1. *The expected uses of monetary evaluation*

1.4.1.1. *The different possible uses of monetary evaluation*

A recent study on the use of the monetary evaluation of ecosystem services lists three main types of possible use: decisive, technical or informative [LAU 13]:

– the economic evaluation of ecosystem services is decisive; when faced with different alternatives, it helps the decider to make a choice in an *ex ante* fashion. The authors distinguish three subcategories. Economic evaluation can help to make trade-offs, notably via a more effective inclusion of ecosystem services in economic analyses. The economic evaluation can also be used as a decision-making tool, promoting an open debate between the parties taking part in a project. Finally, economic evaluation can help to "prioritize conservation efforts" [LAU 13, p. 212];

– economic evaluation can also be used prior to a project or decision to calibrate the instruments that will be used in this framework. This technical use can serve to define compensation levels or to fix prices (e.g. entrance tickets for a park) or taxes on a destructive activity;

– finally, economic evaluation can have an informative end. It can help develop awareness campaigns and used to justify a position or demonstrate the validity of an option, even as a means of verification. Lastly, economic evaluation can serve as an indicator for monitoring.

Other authors focus on the question of how the information is used and on scientific knowledge in public decision making (see, for example, [BAU 12, HER 06, WEI 79]). A new category emerges from this literature, which refers to a strategic use of scientific information. In this case, scientific knowledge (such as, for example, the results of an economic evaluation of services provided by a marine ecosystem) is subject to a "process of perversion of information" [BAU 12, p. 40]. The scientific information can therefore be used to "confirm, or infirm, already acquired knowledge and to legitimate decisions in an *ex post* way" [BAU 12, p. 40]. The scientific information can be used symbolically, in order "to reassure stakeholders by demonstrating the particular importance attached to the objectivation of decisions" [BAU 12, p. 40]. Finally, the supply of new knowledge can be used tactically or procedurally to justify inaction or a delayed strategy.

1.4.1.2. *The use of economic evaluations of ecosystem services*

Despite a large number of applications of economic evaluation methods for ecosystem services, their use as an aid to decision making has had relatively little success.

Concerning the assessment of biodiversity, a study carried out by the OECD concluded that "so far [it] has not achieved the same level of popularity in policymaking as it enjoys in academic circles" [OEC 01, p. 25]. Other studies arrive at the same conclusions (see, for example, [PEA 98, PEA 00, TUR 07]). In 2007, Turner emphasized an "at best 'patchy' take up of CBA by policy makers" [TUR 07, p. 4].

In a study of the academic literature dealing with the economic evaluation of ecosystem services, Laurans *et al.* show that there is after all little information on their use [LAU 13]. Most of the academic papers only briefly referred to the question of use. Of the 313 publications analyzed, only eight relate precisely the role that assessment played in the decision-making process.

Concerning the assessment of marine ecosystem services specifically, a recent study reaches similar conclusions [MAR 13]. On the basis of both academic and gray literature, the authors show that the majority of studies refer to a potential use of economic evaluation in decision making. Only seven references (over 114 studies) report an effective use of the results of evaluation in a decision-making process.

These results, however, only give a partial account of what is at play. With decision-making processes being complex and difficult to study, it is possible that the question of how assessments are used is rarely considered in the literature focusing on monetization alone.

1.4.1.3. *Monetary evaluation: a tool for rationalizing policies or a strategic tool?*

Evaluations of the benefits associated with the restoration of some ecosystem services can also be used in a regulatory framework. As an example, the development of the Water Framework Directive led to the carrying out of cost–benefit analyses on projects for restoring "a good ecological state" (GEE) of land water with a view to knowing if the social costs associated with these projects were not too high in relation to the social benefits expected from these measures. If this was the case, then it could be concluded that there was a problem of a "disproportionate cost", which could trigger a demand for a dispensation[3] to reach the objective fixed by the directive (deferred

3 This dispensation could also be obtained for technical reasons (when no efficient technique is known, or when the preparation of the technique and the action required are too time consuming for the 2015 deadline) and natural reasons (when the reaction time of the milieu necessary for the measures to produce a favorable effect exceeds the 2015 deadline).

from 2021 to 2027 or less ambitious objectives). In France, 710 cost-advantage analyses have been carried out [FEU 14]. In three quarters of these analyses, the costs are higher than the benefits. The authors note that these assessments have above all served to justify decisions based "on other considerations" (politics, acceptability, compromise, etc.) [FEU 14, p. 15]. They underline moreover that the methods for carrying out these assessments are subject to numerous biases.

It therefore seems that the use of economic evaluation in taking decisions is still not clarified. Evaluation appears more as a means of *ex post* justification for decisions that were based on other considerations. This strategic use of assessments risks cause these methods to lose legitimacy. Stakeholder involvement through deliberative or co-constructed approaches for assessments can be a solution to open the debate. This could also enable a greater transparency in economic evaluations.

1.5. A complementary approach: assessing the cost of maintaining ecosystem services

1.5.1. *The principles of assessment*

Here, it is a question of evaluating the costs of environmental policies that aim to meet the objective to preserve the ecological potential that will at once benefit biodiversity for its own sake and also the direct and indirect users of the ecosystem services that it provides [LEV 12].

This method relies on two aspects that distinguish it from the approach above. The first is technical; the method does not have to take into account the uncertainties that exist around the sense-making values of biodiversity and ecosystem services, considering that the restoration of the ecosystems will eventually compensate all the lost well-being that follows a breakdown of the environment – on the condition that the populations that benefit from these restoration measures are the same as those have suffered the loss [ROA 06]. The second is ethical in nature; the method meet a strong sustainability criteria, which depends on the hypothesis that a proportion of the natural capital cannot be substituted by physical or human capital without running a risk of there being a collapse of the ecological system that supports it[4] [EKI 03]. From then on, it is no longer a question of evaluating monetarily the ecosystem services that could be interchangeable, but of assessing the costs of maintaining these ecosystem services.

If these costs can be thought of as the costs of the erosion of biodiversity and ecosystem services, it is because they concern the ensemble of costs that society is willing to pay to maintain a level of biodiversity, under a number of regulations [BAR 09].

1.5.2. *Evaluation of the cost of the degradation of the environment in the directive-framework "Strategy for the marine habitat"*

The directive-framework 2008/56 "Marine Strategy Framework Directive" considers, from its point (2), that "it is evident that the pressures exercised on natural marine resources and the demand for ecosystem marine services are very often too high and that the community should reduce its impact on marine waters" and emphasizes in article 1, line 3, that it is essential that "marine strategies apply an ecosystem-based approach to the management of human activities, enabling it to be guaranteed that the collective pressure resulting from these activities is maintained at levels compatible with the creation of a good ecological state (GES) and to ensure that marine ecosystems' capacity to react to changes caused by nature and by mankind are not compromised, while still permitting sustainable use of marine goods and services by current and future generations."

4 This approach is opposed on principle to weak sustainability, which postulates a perfect substitutability between the different forms of capital [PEA 93]. In effect, the only criteria that counts is that of the economic value furnished by the different form of capital, whatever the relative importance of the latter.

The costs of maintaining the GES of marine waters and services that are associated with it can be described in the following manner [LEV 14]:

– coordination costs correspond to the costs associated with monitoring and surveillance networks for ecosystem services and the sources of impact on these ecosystem services (including the construction of indicators); the running costs of establishments in charge of co-coordinating the protection of marine and coastal ecosystem services (*Agence des aires marines protégées* and *Conservatoire du littoral*); the costs of regulation and control, personnel training and costs for sanctioning the use of protected spaces in particular; and the costs of research, studies, of expertise and assessments of impacts relating to ecosystem marine services;

– costs of positive action correspond to the costs of awareness campaigns, local events and lobbying to limit usage injurious to marine ecosystem services; the costs of basic purchases and putting in place protected marine areas (PMA); and contractual procedures for developing sustainable practices;

– mitigation costs correspond to compensatory measures, ecological restoration programs, the installation of water treatment plants, the extraction of pollutants, etc.

In France, these costs have been estimated for nine themes corresponding to the pressures exercised on marine ecosystem services [LEV 14]. Using these themes, it has been possible to identify legal frameworks of an ecological nature corresponding to the levels of ecosystem services to be maintained (Table 1.7).

If we take the example of the biodiversity loss (Table 1.8), the costs of monitoring and information correspond to the costs associated with monitoring, research, the development of observatories, studies and expertise, coordination between stakeholders in conservation projects, and regulatory procedures. The costs of positive action are those that correspond with strategies for the protection of marine biodiversity. The mitigation costs correspond with the cost of strategies that are carried out after damage to marine biodiversity has occurred.

Themes	Links to GES descriptors	Current legal frameworks
Marine waste	D10 marine waste	OSPAR Convention and Barcelona Convention, legal framework for water treatment, EU Water Framework Directive
Micropollutants	D8 on contaminant concentration levels; D9 on the impacts on health	REACH Directive, legal framework for water treatment and for bathing waters, EU Water Framework Directive
Pathogenic microbial organisms	D9 on the impacts on health	Legal framework for water treatment, for bathing water and the rearing of animals for food, European Directive on water
Oil spills illegal waste/ hydrocarbons	D8 on the concentration levels of contaminants; D9 on the impacts on health	MARPOL, FIPOL, OSPAR Convention and Barcelona Convention
Eutrophisation	D5 on eutrophisation	Nitrate Directive
Impacts on invasive species	D2 on invasive species	Ramsar, CITES, Berne, Bonn, Biodiversity, Barcelona and OMI Conventions
Degradation of biological resources exploited	D3 on the state of exploited stocks	Common Fisheries Policy
Loss of biodiversity seabed integrity, trophic imbalances	D1 on biodiversity; D4 on trophic networks and D6 on seabed integrity	Convention on Biological Diversity, European and French Strategy on Biodiversity
Degradation caused by the introduction of energy into the habitat and by modification of the water regime	D11 on energy and hydrography	Environmental Impact Assessment Directive

Table 1.7. *The themes and ecological references from which the cost of maintaining ecosystem services has been evaluated*

Costs of monitoring and information	Costs of positive action	Costs of attenuation
Monitoring and surveillance network on biodiversity and the sources of impacts on biodiversity (including the construction of indicators)	Awareness campaigns, local events, lobbying to limit usages injurious to marine biodiversity	Compensatory measures
Establishment in charge of coordinating the protection of marine and coastal biodiversity (Agence des AMP and Conservatoire du littoral)	Basic acquisitions by the conservatoire	Restoration and adjustments
Control strategies, personnel training and sanctioning	Creation and management of PMA	
Study, expertise, assessment of impacts	Putting in place of Natura 2000 contracts to develop sustainable practices	
Research on biodiversity		

Table 1.8. *The types of costs measured for the degradation of biodiversity*

By focusing on a specific cost, it is possible to give estimations of costs in detail. These costs for the Channel-North Sea coast are thus detailed in Table 1.9.

Channel-North Sea		
Costs of monitoring and information		
Administration of guardianship and public establishments in charge of protecting the marine habitat	€ 6,147,000	24%
Studies on the impact of gravel extraction	€ 475,000	2%
Professional observatories	€ 2,555,000	10%
Voluntarily run observatories	€ 390,000	2%
Local non-governmental organizations (NGOs)	€ 768,000	3%
Research	€ 15,175,000	59%
Total	€ 25,510,000	100%

Costs of positive action		
Administration of guardianship and public establishments in charge of protecting marine habitat	€ 6,572,000	58%
Main national NGOs	€ 54,000	0%
Protected areas	€ 4,807,000	42%
Total	€ 11,433,000	100%
Costs of attenuation		
Administration of guardianship and public establishments in charge of protecting marine habitat	€ 5,667,000	35%
Protected areas	€ 782,000	5%
Mitigation and compensation for gravel extraction	€ 2,189,000	13%
Seaports	€ 7,763,000	47%
Total	€ 16,401,000	100%

Table 1.9. *Details of the distribution of costs of the degradation of biodiversity for the Channel–North Sea*

The assessment has enabled the calculation for the ensemble of themes and the ensemble of French coasts, an annual maintenance cost for ecosystem services of two billion euros per year [LEV 14].

1.6. Toward multifaceted evaluations of ecosystem services using a spatial approach

1.6.1. *The integrated spatial evaluation of marine and coastal ecosystem services*

The production of ecosystem services on land is today envisaged using a strong spatial specialization, including coastal and marine areas.

Certain coastal zones are specialized in the production of provisioning services with aquaculture and fishing. In these zones, natural variabilities are controlled using mechanization and the use of intensive practices, and this is detrimental to other categories of ecosystem services.

The production zones of cultural services have appeared more recently with increased appreciation of natural sites devoted to walking, recreational fishing, pleasure, diving and snorkeling. In these zones, it is the anthropogenic pressures associated with provisioning services that have been controlled because of regulating access and to a maximization of revenue coming from the tourist industry.

The production zones of regulating services, neglected for a number years, are beginning to be developed in the form, for example, of the restoration of coastal wetlands and protection for the coastline.

This vision, of local areas managed in differentiated ways depending on the category of ecosystem services to be to maximized, appears to be largely outdated. Today, it is expected that a coastal area should be able to maintain professional activities relating to the sea (above all when these activities have a local cultural dimension), this same area should offer the best conditions for the enjoyment of recreational activities relating to natural ecosystems and, finally, it should be managed in a way that guarantees the protection of nature and consequently a healthy environment for the local population. All categories of ecosystem services should thus be taken into account when the question of altering local areas is addressed. Most of the time, this idea of ecosystem service is not explicitly mentioned, but interest in this notion is rightly to be able to make explicit the stakes associated with these management projects from an integrated perspective.

Ecosystem management considers the intra- and interecosystem connections and aims to maintain their capacity to provide multiple services in the long term [MCL 09]. It puts forward a local approach that has political aspects and considers the different actions of management on diverse application scales [LES 10]. This approach can be combined with "marine spatial planning" (MSP) in an "ecosystem services framework" (ESF) to facilitate the sustainable management of marine ecosystems and the services that they provide [GUE 12] (Figure 1.6). MSP represents the decisional approaches that use geospatial information to evaluate how the ocean can be used by people while conserving ecosystem services [CEN 11]. This involves an explicit assessment of compromises in the development of ecosystem services by putting forward a quantitative approach for evaluating the value of the MSP in comparison with planning carried out at random [GUE 12].

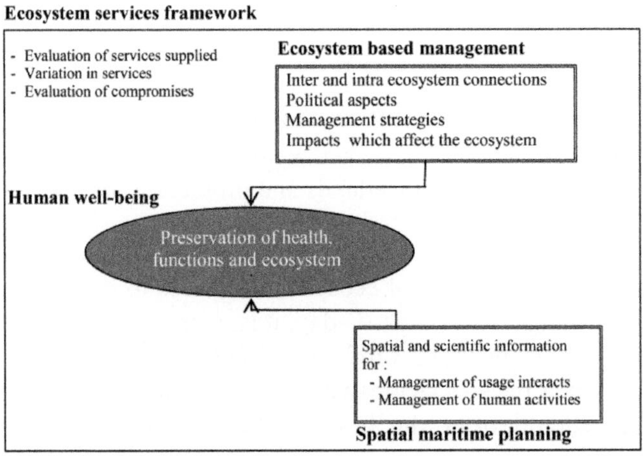

Figure 1.6. *The approach framework for marine and coastal ecosystem services*

An ESF approach is desirable from the point where monetization is no longer useful and therefore is useful in a deliberative context. It requires a good knowledge of the current state and evolution of ecosystem services using several measures, obtained using different models [LEH 13].

The geographical data enable several stakes to be clarified: the distribution and production of ecosystem services, the localization of biodiversity components enabling their production and taking into account of populations and areas benefiting from these services or on the contrary suffering from their reduction, and zones where different usages interact.

A large number of studies have mapped and quantified many ecosystem services and their variations for land habitats [BAI 12, BHA 12, CHA 06, EGO 08, GUL 13, MAE 12, NEL 09, SWE 11]. Nevertheless, studies in marine habitats are much rarer because of the difficulty in obtaining data [GUE 12].

1.6.2. *The spatial integrated assessment of ES and the compromises associated with their development*

The integrated assessment of ES and the compromises associated with their production are shown in Table 1.10. This is inspired by the approach developed by the *Natural Capital Project* [DAI 09].

<table>
<tr><th>Input</th><th>Models</th><th colspan="2">Output</th></tr>
<tr><th>Spatial data</th><th>Direct SE models</th><th>Ecosystem services</th><th>Monetary valorization</th></tr>
<tr><td rowspan="3">Land habitat:
– drainage basin, soil use
Marine habitat:
– habitats, bathymetry, professional and recreational usages

Socioeconomic:
- monitoring of visits and usages, exploitation costs, economic activities</td><td>Carbon
Wave energy
Coastal protection
Recreation
Fishing
Aquaculture
Esthetic quality</td><td rowspan="3">Sequestered carbon
Energy produced
Damages avoided
Number of visitors
Quantity of fishing stocks
Quantity gathered
Number of users per recreational activity</td><td rowspan="3">Value of sequestered carbon
Value of the energy produced
Value of damage avoided
Amount spent on recreational activities
Value of the quantity of fishing stocks
Value of the quantity gathered
Price of housing</td></tr>
<tr><td>Indirect ES models</td></tr>
<tr><td>Water quality
Risks for habitats</td></tr>
<tr><td colspan="4">Participative engagement by stakeholders (managers, users' groups, scientists, etc.) to identify objectives and construct scenarios</td></tr>
</table>

Table 1.10. *The integrated assessment of ES and their compromises [GUE 12]*

The choice of adjustments involving changes in the production of different ecosystem services should be based on trade-offs and compromises since it is impossible to maximize the production of each service simultaneously [BAR 08, HAL 07, TAL 06a]. The identification and representation of ecosystem services using a geographic information system (GIS) can help develop trade-offs. These should be made in collaboration with local stakeholders and be based on a participative and iterative approach [TAL 13b]. The ecosystem services can be mapped and quantified directly, by several models, either in biophysical units (for example the quantity of fish unloaded) or in economic units (for example the value of the quantity fished). Models of indirect ecosystem services, those related to the regulation of water quality or to the risks to habitats, can also be used to evaluate the intermediary effects of management in the flux of services [GUE 12]. Several methodologies among those described above can be used, if this is justified, to determine the services' economic value [LES 10].

After the spatial and quantitative identification of ecosystem services, the use of scenarios can be helpful for better apprehending the dynamics at play and facilitating the compromises to be made. These facilitate an assessment of the impacts on the level of the services provided by the ecosystems due to changes in uses. For example, we can evaluate the quantitative and distributive changes in ecosystem services triggered by an increase in the areas of shellfish exploitation, an increase in the number of fishing licenses or the adoption of more efficient management strategies for different pressures.

1.6.3. *Tools for spatial integrated assessment of ecosystem services*

Many tools are available for modeling marine and coastal ecosystem services. The choice of these tools should be guided by the project's objectives [VIG 11].

The "Multiscale Integrated Models of Ecosystem Services" tool provides models to aid decision making and can integrate maps of changes in soil use [MIM 13]. The models available quantify the effects of land and sea use and their impacts on the level of ecosystem services. These can be applied at global, regional and/or local level.

The tool "artificial intelligence for ecosystem services" [ART 13] takes account of existing information on the different categories of ecosystem services in order to put forward cartographic representations of them using agent-based modeling, GIS and probabilistic Bayesian networks enabling links between variables to be quantified [CHA 10, VIG 11].

The tool "integrated valuation of ecosystem services and trade-offs" (InVEST, [INV 13]) is one family of models for mapping and valuing goods and services furnished by marine and terrestrial areas. InVEST is based on ecological production functions and allows us to understand how changes in the structures, functions and processes of ecosystems impact ecosystem services [DAI 09]. The InVEST models are determinist and based on simplifications of models largely validated by the literature. The user can access the software code and modify it depending on the specific requirements of the project. The first versions of InVEST required the use of the software ArcGIS of ESRI (ArcGIS 2013). The most recent versions also function using open source tools such as Quantum GIS (Quantum GIS 2013) or gvSIG (gvSIG 2013).

An interesting case study on the use of the tool InVEST is one that was carried out for Lemmens Inlet, in British Colombia [GUE 12]. InVEST has been used to evaluate, in a participative context, the impact of different management options on the level of ecosystem services. All the scenarios proposed have explicit spatial (Figure 1.8) and quantitative components (Figure 1.9) enabling discussions with interested parties to be better planned, notably by underlining the necessary compromises, and the consequences of alternative management options.

The three scenarios are as follows:

– *status quo*: no change in the uses or limits of current zones (Figure 1.7(a));

– conservation: restriction of the number of floating homes and aquaculture in the zones close to eel grass. Four floating homes were removed from zones where they were infringing on eel habitats (X in black). Two new concessions of oyster farms in deep water were situated outside sensitive habitat zones (black squares). New kayak routes were proposed (dotted line). Fishing for clams (*Panopea abrupta*) has been forbidden (Figure 1.7(b));

– industrial: five new floating houses have been added (black circles), five new shellfish farming concessions have been added (black squares) and clam fishing is freely permitted (Figure 1.7(c)).

The results of these scenarios on ecosystem services are expressed using simple indicators:

– the risks for ecosystems, which represent the cumulative risk of human activities (floating houses, aquaculture, clam gathering) and coastal habitats (sea eels, soft and hard seabeds), in the light of three management scenarios (Figures 1.7(d)–(f));

– the concentration of fecal coliform bacteria compared to the concentration of floating homes [GUE 12] (Figure 1.7(g)–(i));

– evolutions relative to economic, ecological and social indicators (Figure 1.8).

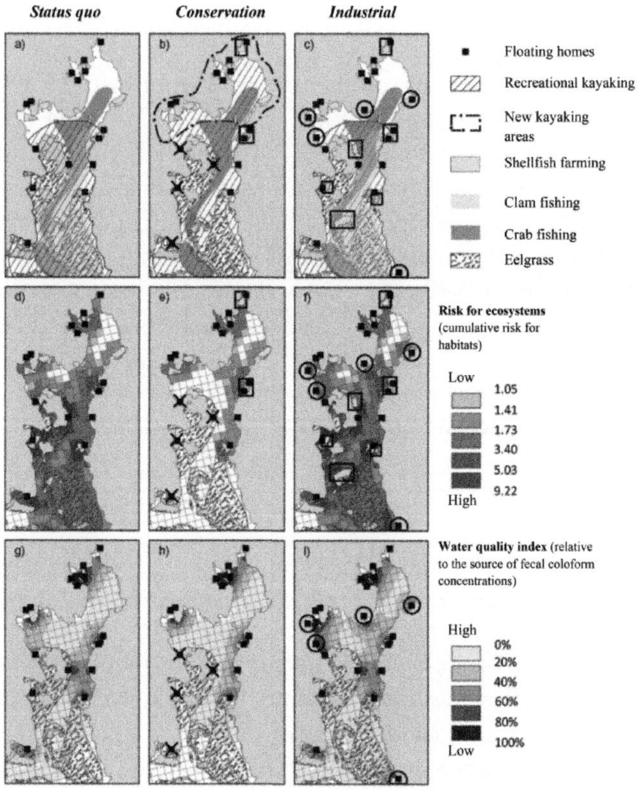

Figure 1.7. *Three management scenarios for Lemmens Inlet (a–c) and some InVEST (d–i) exit strategies [GUE 12] (see color section)*

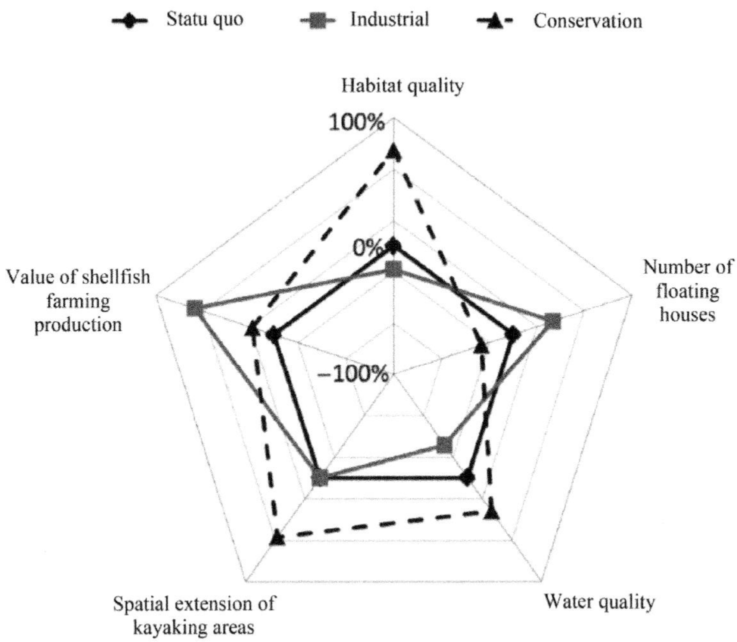

Figure 1.8. *Changes in the production of ecosystem services for three alternative scenarios. The initial levels of ecosystem services are represented by the diamond in continuous black lines. The outward extensions of the base form represent the relative gains and the contractions represent the losses. It is necessary to emphasize that the axis of the water quality is inverted (i.e. the points most removed from the origin have less fecal coliform bacteria and therefore the water quality is higher) [GUE 12]*

1.7. Conclusions

It appears that the notion of ecosystem service can help to improve the management of marine ecosystems by facilitating the trade-offs concerning choices relating to development projects or the conservation of sea shore and marine territory.

To do this, it is necessary to carry out assessments of these ecosystem services. It seems that the monetary approach can be useful when it facilitates an assessment of the costs and benefits associated with direct uses of these services. However, when these uses are too indirect, or indeed non-existent, it does not seem reasonable to seek to propose monetary evaluations of ecosystem services.

This point is reinforced by the report that the use of monetary assessments of marine ecosystem services still seems very limited or is envisaged strategically in the decision-making process.

A complementary assessment, based on multi-criteria approaches to the value of certain ecosystem services, which are difficult to estimate monetarily, can thus appear as a source of supplementary information to facilitate decision making on a local scale. The cartographic supports and use of example scenarios to represent these ecosystem services on different spatial, temporal, and also symbolic scales seem to offer entirely pertinent tools in this regard to accompany the decision-making process, aiming to articulate economic development and the conservation of biodiversity in coastal and marine territories. To this can be added indicators of the cost of the maintenance or restoration of ecosystem services in such a way to balance different levels of effort for different levels of production objectives for ecosystem services on a given local area.

1.8. Bibliography

[APP 04] APPÉRÉ G., "L'évaluation des actifs à usage récréatif : la méthode contingente des coûts de transport", *Revue d'économie régionale & urbaine*, no. 1, pp. 81–106, February 2004.

[ART 13] ARTIFICIAL INTELLIGENCE FOR ECOSYSTEM SERVICES (ARIES), 2013, Available at www.ariesonline.org/.

[BAI 12] BAI Y., ZHENG H., OUYANG Z., *et al.*, "Modeling hydrological ecosystem services and tradeoffs: a case study in Baiyangdian watershed, China", *Environmental Earth Sciences*, December 2012, Available at http://link.springer.com/10.1007/s12665-012-2154-5.

[BAR 08] BARBIER E.B., KOCH E.W., SILLIMAN B.R., *et al.*, "Coastal ecosystem-based management with nonlinear ecological functions and values", *Science*, vol. 319, no. 5861, pp. 321–323, January 2008.

[BAR 09] BARTELMUS P., "The cost of natural capital consumption: accounting for a sustainable world economy", *Ecological Economics*, vol. 68, no. 6, pp. 1850–1857, April 2009.

[BAU 12] BAULER T., "An analytical framework to discuss the usability of (environmental) indicators for policy", *Ecological Indicators*, vol. 17, pp. 38–45, 2012.

[BEA 07] BEAUMONT N.J., AUSTEN M.C., ATKINS J.P., *et al.*, "Identification, definition and quantification of goods and services provided by marine biodiversity: implications for the ecosystem approach", *Marine Pollution Bulletin*, vol. 54, no. 3, pp. 253–265, March 2007.

[BEA 12] BEAUDOIN Y., PENDLETON L., Why Value the Oceans? A Discussion Paper, TEEB, 2012, Available at www.grida.no/files/publications/why-value-the-oceans-discussion-paper-grid-arendal.pdf.

[BHA 12] BHAGABATI N., BARANO T., CONTE M., *et al.*, A green vision for Sumatra: using ecosystem services information to make recommendations for sustainable land use planning at the province and district level, A report by the natural capital project, WWF-US, and WWF-Indonesia, 2012. Available at: www.ncp-dev.

[BON 07] BONTEMS P., ROTILLON G., *L'économie de l'environnement*, 3rd ed., La Découverte, Paris, 2007.

[CAS 05] CASTELLA J.C., "Participatory simulation of land-use changes in the northern mountains of Vietnam: the combined use of an agent-based model, a role-playing game, and a geographic information system", *Ecology and Society*, vol. 10, no. 1, 2005.

[CEN 09] CENTRE D'ANALYSE STRATÉGIQUE, Approche économique de la biodiversité et des services liés aux écosystèmes, La Documentation française, Paris, 2009.

[CEN 11] CENTER FOR OCEAN SOLUTIONS, Decision-guide: selecting decision-support tools for marine spatial planning, 2011, Available at http://centerforoceansolutions.org/sites/default/files/cos_mspguide_6.pdf.

[CHA 06] CHAN K.M.A., SHAW M.R., CAMERON D.R. *et al.*, "Conservation planning for ecosystem services", *PLoS Biol*, vol. 4, no. 11, p. e379, October 2006.

[CHA 10] CHAN K.M.A., RUCKELSHAUS M., Characterizing changes in marine ecosystem services, F1000 biology reports, July 2010, Available at: www.f1000biology.com/reports/10.3410/B2-54.

[CLA 73] CLARK C.W., "The economics of overexploitation", *Science*, vol. 181, no. 4100, pp. 630–634, August 1973.

[CON 04] CONVENTION SUR LA DIVERSITÉ BIOLOGIQUE, Décisions adoptées par la conférence des parties à la convention sur la diversité biologique à sa septième réunion – Annexe, p. 375, 2004, Available at: www.cbd.int/doc/decisions/cop-07/full/cop-07-dec-fr.pdf.

[CON 98] CONVENTION SUR LA DIVERSITÉ BIOLOGIQUE, COP 4 Décision IV/10, 1998, Available at: www.cbd.int/decision/cop/default.shtml?id=7133.

[COS 97] COSTANZA R., D'ARGE R., DE GROOT R., *et al.*, "The value of the world's ecosystem services and natural capital", *Nature*, vol. 387, no. 6630, pp. 253–260, May 1997.

[COU 06] COURCHAMP F., ANGULO E., RIVALAN P., *et al.*, "Rarity value and species extinction: the anthropogenic allee effect", *PLoS Biol*, vol. 4, no. 12, pp. e415, November 2006.

[DAI 97] DAILY G.C., *Nature's Services: Societal Dependence on Natural Ecosystems*, Island Press, Washington, 1997.

[DAI 08] DAILY G.C., MATSON P.A., "Ecosystem services: from theory to implementation", *Proceedings of the National Academy of Sciences*, vol. 105, no. 28, pp. 9455–9456, July 2008.

[DAI 09] DAILY G.C., POLASKY S., GOLDSTEIN J., *et al.*, "Ecosystem services in decision making: time to deliver", *Frontiers in Ecology and the Environment*, vol. 7, no. 1, pp. 21–28, February 2009.

[DIA 94] DIAMOND P.A., HAUSMAN J.A., "Contingent valuation: is some number better than no number?", *Journal of Economic Perspectives*, vol. 8, no. 4, pp. 45–64, Autumn 1994.

[DE 12] DE GROOT R., BRANDER L., VAN DER PLOEG S., *et al.*, "Global estimates of the value of ecosystems and their services in monetary units", *Ecosystem Services*, vol. 1, no. 1, pp. 50–61, July 2012.

[DE 13] DE GROOT R., BLIGNAUT J., VAN DER PLOEG S., *et al.*, "Benefits of investing in ecosystem restoration", *Conservation Biology*, vol. 27, no. 6, pp. 1286–1293, December 2013.

[DUA 00] DUARTE C.M., "Marine biodiversity and ecosystem services: an elusive link", *Journal of Experimental Marine Biology and Ecology*, vol. 250, nos. 1–2, pp. 117–131, July 2000.

[EGO 08] EGOH B., REYERS B., ROUGET M., *et al.*, "Mapping ecosystem services for planning and management", *Agriculture, Ecosystems & Environment*, vol. 127, nos. 1–2, pp. 135–140, August 2008.

[EHR 83] EHRLICH P.R., MOONEY H.A., "Extinction, substitution, and ecosystem services", *BioScience*, vol. 33, no. 4, pp. 248–254, April 1983.

[EKI 03] EKINS P., SIMON S., DEUTSCH L. *et al.*, "A framework for the practical application of the concepts of critical natural capital and strong sustainability", *Ecological Economics*, vol. 44, nos. 2–3, pp. 165–185, March 2003.

[ENV 09] ENVIRONMENTAL PROTECTION AGENCY (EPA), Valuing the Protection of Ecological Systems and Services, Report, Environmental Protection Agency, Washington, 2009.

[ESR 13] ESRI, ArcGIS, 2013, Available at: www.esri.com.

[EUR 13] EUROPEAN COMMISSION, Proposal for a directive of the European Parliament and of the council establishing a framework for maritime spatial planning and integrated coastal management, Brussels, p. 35, 2013, Available at http://eur-lex.europa.eu/LexUriServ/LexUriServ.do?uri=COM:2013:0133:FIN:EN:PDF.

[FEU 14] FEUILLETTE S., LEVREL H., BLANQUART S., *et al.*, "L'évaluation monétaire de la nature à l'épreuve de la gestion de l'eau. Retour d'expériences des agences de l'eau et mise en discussion", *Natures, Sciences, Sociétés*, in press, 2014.

[GAU 08] GAULT A., MEINARD Y., COURCHAMP F., "Consumers' taste for rarity drives sturgeons to extinction", *Conservation Letters*, vol. 1, no. 5, pp. 199–207, 2008.

[GUE 12] GUERRY A.D., RUCKELSHAUS M.H., ARKEMA K.K., *et al.*, "Modeling benefits from nature: using ecosystem services to inform coastal and marine spatial planning", *International Journal of Biodiversity Science, Ecosystem Services & Management*, vol. 8, nos. 1–2, pp. 107–121, January 2012.

[GUL 13] GULICKX M.M.C., VERBURG P.H., STOORVOGEL J.J., *et al.*, "Mapping landscape services: a case study in a multifunctional rural landscape in The Netherlands", *Ecological Indicators*, vol. 24, pp. 273–283, January 2013.

[GVS 13] GVSIG, "GvSIG", 2013, Available at: www.gvsig.org/.

[HAL 07] HALPERN B.S., SELKOE K.A., MICHELI F., *et al.*, "Evaluating and ranking the vulnerability of global marine ecosystems to anthropogenic threats", *Conservation Biology: the Journal of the Society for Conservation Biology*, vol. 21, no. 5, pp. 1301–1315, October 2007.

[HEA 00] HEAL G.M., *Nature and the Marketplace: Capturing the Value of Ecosystem Services*, Island Press, Washington, 2000.

[HEZ 06] HEZRI A.A., DOVERS S.R., "Sustainability indicators, policy and governance: issues for ecological economics", *Ecological Economics*, vol. 60, pp. 86–99, 2006.

[HOL 99] HOLMLUND C.M., HAMMER M., "Ecosystem services generated by fish populations", *Texte*, vol. 29, pp. 253–268, 1999.

[INV 13] INVEST, "Integrated valuation of environmental services and tradeoffs", 2013. Available at www.naturalcapitalproject.org/.

[JAC 01] JACKSON J.B.C., KIRBY M.X., BERGER W.H., *et al.*, "Historical overfishing and the recent collapse of coastal ecosystems", *Science*, vol. 293, no. 5530, pp. 629–637, July 2001.

[KAH 92] KAHNEMAN D., KNETSCH J.L., "Valuing public goods: the purchase of moral satisfaction", *Journal of Environmental Economics and Management*, vol. 22, no. 1, pp. 57–70, January 1992.

[KAH 94] KAHNEMAN D., RITOV I., "Determinants of stated willingness to pay for public goods: a study in the headline method", *Journal of Risk and Uncertainty*, vol. 9, no. 1, pp. 5–37, 1994.

[KAI 11] KAISER M.J., ATTRILL M.J., JENNINGS S., *et al.*, *Marine Ecology: Processes, Systems, and Impacts*, Oxford University Press, Oxford, 2011.

[KAL 10] KALAYDJIAN R., *Données économiques maritimes françaises 2009*, Editions Quae, Versailles, 2010.

[KRE 05] KREMEN C., "Managing ecosystem services: what do we need to know about their ecology?", *Ecology Letters*, vol. 8, no. 5, pp. 468–479, 2005.

[LAF 07] LAFFITTE P., SAUNIER C., Les apports de la science et de la technologie au développement durable, tome II: La biodiversité: l'autre choc? l'autre chance? – Sénat, p. 192, 2007, Available at: www.senat.fr/notice-rapport/2007/r07-131-notice.html.

[LAU 13] LAURANS Y., RANKOVIC A., BILLÉ R., *et al.*, "Use of ecosystem services economic valuation for decision making: questioning a literature blindspot", *Journal of Environmental Management*, vol. 119, pp. 208–219, 2013.

[LEH 13] LEH M.D.K., MATLOCK M.D., CUMMINGS E.C., *et al.*, "Quantifying and mapping multiple ecosystem services change in West Africa", *Agriculture, Ecosystems & Environment*, vol. 165, pp. 6–18, January 2013.

[LES 10] LESTER S.E., MCLEOD K.L., TALLIS H., *et al.*, "Science in support of ecosystem-based management for the US West Coast and beyond", *Biological Conservation*, vol. 143, no. 3, pp. 576–587, March 2010.

[LEV 12] LEVREL H., HAY J., BAS A., *et al.*, "Coût d'opportunité versus coût du maintien des potentialités écologiques: deux indicateurs économiques pour mesurer les coûts de l'érosion de la biodiversité", *Natures, Sciences, Sociétés*, vol. 20, pp. 1–14, 2012.

[LEV 14] LEVREL H., JACOB C., BAILLY D., *et al.*, "The maintenance costs of marine natural capital: a case study from the initial assessment of the Marine Strategy Framework Directive in France", *Marine Policy*, vol. 49, pp. 37–47, 2014.

[LIQ 13] LIQUETE C., PIRODDI C., DRAKOU E.G., *et al.*, "Current status and future prospects for the assessment of marine and coastal ecosystem services: a systematic review", *PLoS ONE*, vol. 8, no. 7, p. e67737, July 2013.

[MAE 12] MAES J., EGOH B., WILLEMEN L., *et al.*, "Mapping ecosystem services for policy support and decision making in the European Union", *Ecosystem Services*, vol. 1, no. 1, pp. 31–39, July 2012.

[MAR 13] MARCONE O., MONGRUEL R., "Looking beyond academic application of environmental valuation: what practical uses of economic valuation in decision-making processes? The Case of Marine Environment", *10th biennal conference of the European Society for Ecological Economics*, Lille, p. 22, 8–21 June 2013.

[MCL 09] MCLEOD K., LESLIE H., *Ecosystem-Based Management for the Oceans*, Island Press, Washington, 2009.

[MIL 05] MILLENIUM ECOSYSTEM ASSESSMENT (MEA), *Ecosystem and Human Well-Being: Synthesis*, Island Press, Washington, 2005.

[MIL 11] MILANESI J., "Une histoire de la méthode d'évaluation contingente", *Genèses*, vol. 84, pp. 6–24, October 2011.

[MIM 13] MIMES, Multiscale Integrated Models of Ecosystem Services, MIMES, 2013, Available at: www.ebmtools.org/.

[MON 13] MONGRUEL R., LEVREL H., "The institutional foundations of the ecosystem services approach", *10th biennal conference of the European Society for Ecological Economics*, Lille, p. 14, 8–21, June 2013.

[NAE 09] NAEEM S., *Biodiversity, Ecosystem Functioning, and Human Wellbeing: An Ecological and Economic Perspective*, Oxford University Press, 2009.

[NEL 09] NELSON E., MENDOZA G., REGETZ J., *et al.*, "Modeling multiple ecosystem services, biodiversity conservation, commodity production, and tradeoffs at landscape scales", *Frontiers in Ecology and the Environment*, vol. 7, no. 1, pp. 4–11, February 2009.

[NOR 98] NORGAARD R.B., BODE C., "Next, the value of god, and other reactions", *Ecological Economics*, vol. 25, no. 1, pp. 37–39, April 1998.

[OCO 99] O'CONNOR M., SPASH C.L., *Valuation and the Environment. Theory, Methods and Practices*, Edward Elgar, Cheltenham, 1999.

[OEC 01] OECD, *Valuation of Biodiversity Benefits. Selected Studies*, OECD Publications, Paris, 2001.

[OEC 02] OECD, *Manuel d'évaluation de la biodiversité: Guide à l'intention des décideurs*, OECD Publishing, Paris, 2002.

[OEC 12] OECD, *OECD Environmental Outlook to 2050*, OECD Publishing, 2012.

[PAY 93] PAYNE J.W., "Where do the numbers come from? How people respond to contingent valuation questions?", *Contributions to Economic Analysis*, pp. 271–303, North Holland, July 1993.

[PEA 00] PEARCE D.W., SECCOMBE-HETT T., "Economic valuation and environmental decision-making in Europe", *Environmental Science & Technology*, vol. 34, pp. 1419–1425, 2000.

[PEA 93] PEARCE D.W., ATKINSON G.D., "Capital theory and the measurement of sustainable development: an indicator of 'weak' sustainability", *Ecological Economics*, vol. 8, no. 2, pp. 103–108, October 1993.

[PEA 98] PEARCE D.W., "Environmental appraisal and environmental policy in the European Union", *Environmental and Resource Economics*, vol. 11, nos. 3–4, pp. 489–501, 1998.

[PET 03] PETERSON G.D., CUMMING G.S., CARPENTER S.R., "Planificación de un Escenario: una Herramienta para la Conservación en un Mundo Incierto", *Conservation Biology*, vol. 17, no. 2, pp. 358–366, 2003.

[QGI 13] QGIS, Quantum GIS, 2013, Available at: www.qgis.org.

[ROA 06] ROACH B., WADE W.W., "Policy evaluation of natural resource injuries using habitat equivalency analysis", *Ecological Economics*, vol. 58, no. 2, pp. 421–433, June 2006.

[RON 07] RÖNNBÄCK P., KAUTSKY N., PIHL L., *et al.*, "Ecosystem goods and services from Swedish coastal habitats: identification, valuation, and implications of ecosystem shifts", *Ambio*, vol. 36, no. 7, pp. 534–544, November 2007.

[RUF 09] RUFFO S., KAREIVA P.M., "Using science to assign value to nature", *Frontiers in Ecology and the Environment*, vol. 7, no. 1, p. 3, February 2009.

[SOL 04] SOLAN M., CARDINALE B.J., DOWNING A.L., *et al.*, "Extinction and Ecosystem Function in the Marine Benthos", *Science*, vol. 306, no. 5699, pp. 1177–1180, December 2004.

[SOM 13] SOMDA J., AWAISS A., *Evaluation économique des fonctions et services écologiques des écosystèmes naturels: Guide d'utilisation de méthodes simples*, UICN, Ouagadougou, 2013.

[SWA 94] SWANSON T.M., "The economics of extinction revisited and revised: a generalised framework for the analysis of the problems of endangered species and biodiversity losses", *Oxford Economic Papers*, vol. 46, pp. 800–821, 1994.

[SWE 11] SWETNAM R.D., FISHER B., MBILINYI B.P., *et al.*, "Mapping socio-economic scenarios of land cover change: a GIS method to enable ecosystem service modeling", *Journal of Environmental Management*, vol. 92, no. 3, pp. 563–574, March 2011.

[TAL 06a] TALLIS H., KAREIVA P., "Shaping global environmental decisions using socio-ecological models", *Trends in Ecology & Evolution*, vol. 21, no. 10, pp. 562–568, October 2006.

[TAL 06b] TALLIS H., RICKETS T., GUERRY A., *et al.*, InVEST 2.5.6 user's guide: integrated valuation of environmental services and tradeoffs, The Natural Capital Project, 2013. Available at http://ncp-dev.stanford.edu/~dataportal/invest-releases/documentation/current_release/.

[TEE 10] TEEB, Economie des écosystèmes et de la biodiversité, Rapport pour les entreprises, Summary, 2010.

[TEE 13] TEEB, TEEB for oceans and coasts, 2013. Available at www.teebweb.org/wp-content/uploads/2013/04/2013-Oceans-and-Coasts-Flyer.pdf.

[TRA 08] TRAVERS M., NASSIRI A., APPÉRÉ G., *et al.*, "Evaluation des bénéfices environnementaux par la méthode des prix hédonistes: une application au cas du littoral", *Economie & Prévision*, vol. 185, pp. 47–62, December 2008.

[TUR 07] TURNER R., "Limits to CBA in UK and European environmental policy: retrospects and future prospects", *Environmental & Resource Economics*, vol. 37, no. 1, pp. 253–269, 2007.

[UNE 06] UNEP, Marine and Coastal Ecosystems and Human Well-Being: Synthesis, United Nations Environment Programme, Nairobi, Kenya, 2006.

[UNE 11] UNEP-WCMC, "Marine and coastal ecosystem services. Valuation methods and their practical application", *NEP-WCMC Biodiversity Series*, p. 46, 2011.

[UNE 12] UNEP, FAO, IMO *et al.*, Green economy in a blue world, Synthesis report, 2012, Available at www.unep.org/pdf/green_economy_blue.pdf.

[UNE 13] UNEP, Millennium ecosystem assessment 2013. Available at www.unep.org/maweb/fr/About.aspx.

[VAN 02] VANOLI A., *Une histoire de la comptabilité nationale*, La Découverte, Paris, 2002.

[VIG 11] VIGERSTOL K.L., AUKEMA J.E., "A comparison of tools for modeling freshwater ecosystem services", *Journal of Environmental Management*, vol. 92, no. 10, pp. 2403–2409, October 2011.

[WEI 79] WEISS C.H., "The many meanings of research utilization", *Pubic Administration Review*, vol. 39, no. 5, pp. 426–431, 1979.

[WOR 06] WORM B., BARBIER E.B., BEAUMONT N., *et al.*, "Impacts of biodiversity loss on ocean ecosystem services", *Science*, vol. 314, no. 5800, pp. 787–790, March 2006.

2

Fisheries and Aquaculture Sustainability

2.1. Sustainability and responsibility of provisioning: learning the lessons from overfishing

This chapter does not hide its ambitions to use the subject of fishery to propose a transdisciplinary synthesis of theories and of management concepts, which seem to us to be the most developed, to evaluate the sustainability of human activities. The French-speaking community uses the terms durability (*durabilité*) and sustainability (*soutenabilité*) in French and some latin languages to designate the properties grouped under the dedicated concept of *Sustainable Development*. In French, the term "soutenable" becomes a professional and academic term whereas the term "durable" is that of the political world and of the general public (see Appendix 1). Furthermore, this discussion will consider the term of responsibility regarding the transformation of the biosphere by the human species.

Chapter written by Sylvestre VOISIN and Pierre FRÉON.

2.1.1. *Introduction: challenges and summary of key points*

In this introduction, we wish to present a general view of the ambitions of this chapter and a summary of the ideas that are developed. The uninformed reader may perceive the most advanced paths to a more responsible fisheries management. Informed readers will be able to evaluate an attempt of interdisciplinary demonstration that the sustainability of fishing and marine resources depends on the sustainability of the anthropogenic exploitation of the ocean (i.e. by human societies, both local and diverse, but globally connected by their means and their economy) [BAC 10, GIA 98].

We have noticed that the current ecological theories have difficulty in justifying the calculations of energy and matter flows that take fishing exploitation from the ocean to the plate of the consumer. The unification of conventional naturalist approaches and more recent approaches in industrial ecology (IE) will be necessary to clarify the conventions about energy-efficiency to apply to the whole sector. We are developing a generalized ecology (GE) approach in order to propose a new model according to which fishing exploitation transfers biomass, collected by artificial means, from an exploited ecosystem (ocean) toward a terrestrial exploiting system (the human societies or their economies) whose metabolism and evolution are not the same [FRO 08]. Table 2.1 and Appendix 2 compare this approach to the previous ones. We want to show (in theory and with a case study) how evaluation methods of environmental, economic and social impacts of human activities arising from "lifecycle analysis", or LCA, can constitute an accounting framework of the sustainability of fishing and of aquaculture supply chains [EUR 10, HEN 12, ISO 06, JOL 10, PEL 06, VAL 11]. These evaluations are performed on a perimeter expanded to human societies "from the ocean to the plate" [FRE 10]. They promote consistency and operationality of the management parameters identified under the banner of sustainable development, of sustainable economy and more recently of the social responsibility of organizations.

We therefore propose to make the transition from a logic of fishery (and aquaculture) to a logic of human societies provisioning, more significant from the energy-efficiency point of view, in order to evaluate the sustainability and the responsibilities across the supply chains[1]. The point is to legitimize the approaches of sustainable exploitation with regard to the current market logics resulting in an irresponsible exploitation. Both relate to the renewal of resources, as well as direct and indirect collateral damage caused to marine and terrestrial ecosystems by the human economy as a whole. The approach involves reorganizing these provisioning, or supply, chains by moving away from the logic of means (production), to raise the issue concerning the purpose of consumption and arbitrations in order to choose the most effective and sustainable prospects [FAO 99, FAO 12, OEC 07, PAU 02, VAL 07, WOR 08].

Management Approach	Founding authors	Keywords	Management criteria	Key concepts
Initial	[HJO 14] [GRA 35] [SCH 57] [GOR 54] [BEV 57] [RIC 58]	MSY Monospecific Monodisciplinary	Captures, efforts, more rarely classic economic optimum	Monospecific quotas
Ecosystem (EAF) and (EBFM) bio-economic	[GAR 03] [CLA 85]	Ecosystem MSY Multi-species Multidisciplinary	Local ecological and economic indicators, sometimes environmental (bio-eco environmental models), catches and effort adapted to the biology of the species	Marine Protected Areas (MPAS) and strict quotas

1 In this chapter the term “supply chain” is used as a generic term for value chain or provisioning chain. A supply chain is here a chain of stakeholders or agents (individuals or firms) that operate to catch, transform and distribute seafood or products to the end consumer.

Industrial ecology and ecological economics	[AYR 94a] [BRO 03] [DAL 11] [GRE 10] [COS 02]	Extracted biomass flow, energetic and fleet matters weight of upstream chains	Material and energy balances of fleets and supply chains of transformation, balances of environmental, economical and social impacts, responsible consumption	Equipement and processes LCA, supply chains LCSA, industrial metabolism and human domestic
Generalized ecology and systems panarchical approach	[FOL 98] [FRO 08] [GIA 07] [GUN 02] [GER 12] and this chapter	Food supply and consumption, metabolism of human societies, adaptive cycles	Unified eco-energetic, costs and overall performance of provisioning, sustainable consumption and responsible, eco-social approaches and prosperity heterodox economy	Energies of biomass production, human auxiliary energy, exploitation of an ecosystem by another, panarchy cycles and complex regulations

Table 2.1. *Comparison of the approaches and the scientific paradigms of fishery activity*

The result of these evaluations require a political interpretation, which can be facilitated by the use of contrasting scenarios, with the objective of supporting the decision process. This exercise helps to visualize possible approaches with which to supply human societies in seafood products, or other auxiliary uses. The evaluation of these scenarios follows an integrated approach which allows at any time the semi-quantitative balance evaluation of ecological, environmental, economic and social (or political) stock properties of the supply chains, globally or by segment. It is then possible to test the parameters of their sustainability, or non-sustainability, according to the specificities of the various regions of the globe, of the fishing fleets involved and the use of marine aquaculture [VAL 11].

It is difficult to assess the question of a tendency to overfish and of the total exploitation of the oceans on a global scale [FAO 12]. This situation demonstrates the unsustainability of the initial fisheries approach (see section 2.1.2.1) and supply chains that they cater for, except in some developing countries that can invest in stock assessment and monitoring, and can afford a low exploitation rate. The former approach is still dominant today. It is based on the paradigm of defining stocks of exploitable marine species[2] independently of each other (monospecific approach) on purely demographic considerations (elementary approach of the biology of species) and on the assumption that we can assign them to existing areas of fisheries, often exploited by a single fleet. However, fisheries research has strongly progressed to demonstrate the theoretical and practical limits, by proposing an ecosystem-based fisheries management (EBFM) or the related ecosystem approach to fisheries (EAF), more complex than the simple stock management. The few recent applications of the EAF show that progress is possible, but today is only achieved by the richer countries [WOR 12] who can afford to pay the price of regulation and for which the provisioning of seafood is not critical.

Several approaches defend the position that the sustainability of fishery and marine aquaculture activities should be evaluated, while also considering the economic opportunities, formal or not, of fishing products [WOR 08]. Sustainability involves both the exploiting humans, who equipt the fleets (small-scale as well as industrial) and consume with little concern the products extracted from the ocean, as well as the exploited ecosystems themselves, whose vulnerability has only been perceived recent. However, fishery research has a tendency to stay centered on the upstream segment of human societies provisioning, following a logic of dominant production. Yet, the environmental, ecological and even social or societal sustainability of fishery and aquaculture is better evaluated within the whole worldwide anthropogenic system (anthroposphere), according to a logic of

2 A stock will be defined as a management unit on an artificially delimited marine space, within of a population that could be more extensive. Although exchanges of genes can take place between stocks, they must be sufficiently limited in order to prevent any significant interference on the population dynamics on timescales smaller than the interdecadal scale.

consumption and productivity for human development, with regards to the indicators of human development of international institutions.

Our demonstration of a new sustainability framework, accounting for overall global performance, will focus especially on the fishing activity but can also easily integrate marine aquaculture. The globalization of performances implies that account must be taken of the fact that the fishing, as well as the aquaculture, supply chains agglomerate with other supply chains. It concerns food chains, in terms of biomass flows, and materials and energy supply chains, with respect to equipments and logistics. The whole integrates the metabolism of human societies, considered as a fully-fledged system, as illustrated by our case study. We will see how an accounting sustainability system in global performance allows for allocating costs or contributions to the metabolism in proportion of the weight of the supply chains being considered. This chapter is structured in three parts.

In the first part (section 2.1), which continues after these introductory remarks (section 2.1.1), we assume the observation that the initial approach to fishery is challenged by the undeniable facts of overfishing (section 2.1.2). We will summarize the contributions of relatively new sciences, or of advanced academic developments, that we mobilize for our demonstration on GE and on the assessment of sustainability (section 2.1.3). We will then try to formulate a new approach for the sustainability of fisheries and marine aquaculture provisioning (section 2.1.4).

In the second part (section 2.2), we will put forward state of the art trends (section 2.2.1) and a demonstrative case study (section 2.2.2) of multi-criteria evaluation methods of the sustainability of fisheries and aquaculture systems. Then, we will present an analysis of the relevance of the concepts of EAF/EBFM, of bioeconomy, and of the sustainable and responsible consumption of sea products (section 2.2.3) considering the potential of sustainability assessment methods.

The third part (section 2.3) deals with the political and managerial evaluation of the sustainability and responsibility of fisheries and marine aquaculture exploitation by global performance. To illustrate this approach, we detail a few large contrasting scenarios to describe the scope of the possible political debates, to be adapted according to the regions of the world. The use of the method of contrasting scenarios is a powerful pedagogy tool for decision support. We will outline a few principles of proper procedure of these evaluations to effectively translate the scientific knowledge into indicators that can be assimilated by political debates. We will conclude by summarizing what seems to characterize the integrated, sustainable and responsible approaches for the management of fisheries and aquaculture (2.3.3).

2.1.2. *The initial fisheries approach challenged by the complexity of the ocean ecosystem and the failures of fisheries regulation*

2.1.2.1. *The initial approach*

An approach to fishery, that we will call "initial", established, since the 1950s [HIL 92], that fishing should and could respect the biology of the exploited wild species in order not to cause the extinction of populations (see Chapter 4 of [MON 14a]). This monospecific approach has been supported by states because it is simple and consistent with the specialization of some fleets and their trading opportunities. It has generated a rapid development of measurement and, direct (e.g. hydroacoustic campaigns) or indirect (e.g. demographic sampling), scientific monitoring tools which are often expensive. These observation and modeling tools have enabled to determine the demographic and biological productivity parameters of the exploited species, and to deduce the criteria for the regulation of fisheries. The fisheries ecologists have proposed quite quickly the key concept of "maximum sustainable yield", or MSY, which assumes that below a maximum level of abundance (maximal biotic capacity), any growing population produces a "surplus" which can be levied without compromising the level of stock catches. This MSY is associated with a maximum sustainable effort, or MSE, which attempts to quantify the capacity of fleets (see Chapter 2 of [MON 14b]). These models, attractive and widely accessible, are determined for each species, which promotes a monospecific approach or management by categories of "resources", such as traded on markets after the unloading of the fishery. The central notion of the MSY has allowed to adapt models to some of the biological subtleties, finding a compromise within the mechanisms of the economy in an attempt to correct the risks of overfishing and to establish regulations.

The application of the initial approach in Europe, since the beginning of the industrial fishing era in the first half of the 20th Century, did not prevent the overexploitation of fish species [CUR 08, MUL 05, RAM 02]. The phenomenon of widespread of overfishing and of full-scale exploitation of all the seas since the 1990s [FAO 12], is testament of the fact that the fishing supply chains of the industrial era have generally not been able to organize themselves to adjust the pressure of fishery to the fluctuations in the abundance of fish or of the other coveted "sea products". They have refused, and still largely refuse, the application of the initial, yet simple, approach, which they consider challenges their secular business and its myths. The vastness of the ocean and the deficiencies of the national and international maritime law have long perpetuated a regime of free access. The latter has often been facilitated by the allocation of state subsidies causing the distortion of fleets cost across different countries (see Chapter 3). These practices have almost always resulted in the collapse of the exploitable stocks, or even in their extinction. Up until the realization of the negative effects of fishing in the 1990s, beliefs and ignorance maintained the myth of a generous sea, of miraculous fisheries and of the inexhaustible nature of the ocean, as promoted by Huxley [HUX 84]. Overall, even in the vicinity of the coasts, the capacities of states to defend rational policies of access to the resource remain limited (see Chapter 3, although to nuance according to the development level of the exploiting countries [PIT 13]). Very often, the agreements on quotas or on fishing periods do not succeed, or the agreements take the form of a not very realistic political compromise given the state of the resource. The regulations can also be circumvented or infringed, accordingly due to the lack of control means from the authorities. Finally, the resource can move or fluctuate beyond the forecast of the assessors and beyond the definition of fishing areas. The numerous examples of fisheries with collapsed resources have not slowed down the arms race of the fleets which, in addition, complicates the statistical monitoring of fishing supply chains. The balance of the applications of the initial fisheries approach thus defined brings forward four main points detailed below.

2.1.2.2. *Limits of the monospecific MSY and the derived indicators*

The concept of maximum biotic capacity of the environment corresponds to a quite classical (static) vision of ecology and of the productivity of ecosystems. In reality, the marine species can be highly mobile and the external forcing factors (natural or anthropogenic) highly unpredictable. As a result, the biotic capacity of a fishing area, as well as the geographical delimitation of stocks and ecosystems are concepts difficult to apply on the ground. It is shown that, for a lot of resources fluctuating according to interdecadal, or even decadal, pseudo-cycles, the exploitation can only be optimal in the light of the theoretical requirements for the renewal of the resource if the annual catch is located well below the MSY [BOU 87, FRE 05, FRE 08, THE 98]. In addition, the rational management of stocks assumes that fleets or exploitation systems become perfectly controllable such that the "tragedy of commons" does not impose itself, being condemned to disappear if they are freely accessible and without owner, according to Hardin [BAD 98, HAR 68].

Fisheries economists therefore propose the determination of a bioeconomic equilibrium, often achieved at the cost of a fishing discipline somehow unacceptable to the exploiters. The corresponding models are based on several hypotheses, of a "freely accessible" renewable resource, with a "single-owner", "structured by age classes", in static, in dynamics, with random variations, etc. (see Chapter 3, and also Chapter 2 of [MON 14b]). This approach of the regulations, initially inspired by biologists, has served as the basis for the economic theory of the exploitation of renewable living natural resources [FAU 95, GOR 54, PER 96, TIE 00]. This approach created an economic constraint which has stimulated some very innovative organization modes of fisheries by playing with the effects of the reduction of volumes placed on the market, on the prices and profitability of the extraction. Generally, the best regulations are observed on the resources easy to locate and to assess, on products with high value and high demand. These regulated fisheries tend toward an exploitation that can be described as extensive and which seems quite coherent with the principle of adapting itself to the productivities of non-domesticated ecosystems (model of hunting-gathering).

The difficulties in defining and enforcing regulations on the common goods that the renewable living resource of the oceans represents are at the origin of a major criticism of the organization of seafood supply chains [GUI 08, MUL 05, PAU 02]. The initial approach of fish exploitation refers, however, implicitly to the neoclassical economic school to define the mechanisms operating upon the creation of prices and of the economic opportunities of fishing products (value chains) (to get a better idea, see Chapter 3). These opportunities are very varied and as a result do not condition the fleets according to the same socio-economic organization modes. They are either small local food supply chains, which are eventually non-monetarized, or either bigger regional or world supply chains, monetarized and more or less industrialized [ALD 08, FAO 12, FRE 83, FRE 13]. However, the calculations on stocks and on the exploitability of the resources being uncertain, notions apparently solid about the fisheries economy proved impossible to manage in practice (rent or pseudo-rent theory; see Chapter 3 section 3.2.1.4). In addition, the rights of access to resources are difficult to establish in the absence of marine delimitations, except in the immediate proximity of the coasts [BOU 87].

Finally, and in part due to the increasingly more frequent cases of overfishing, research has established the dependence of exploited species to trophic or food webs that relate to and structure the ecosystem [GAR 12]. Therefore, the accessory catches and discard at sea of non-marketable catches, which represent sometimes more than 50% of mass catches before the sorting on the boat, are the indirect components of fishing which impact on the ecosystem have been demonstrated (imbalances in the natural food chains [GRE 08]). Finally, fisheries seems to have an impact on all the species of the trophic chain and on the general parameters of marine ecosystems, according to rules which are beginning emerge from the multiple case studies across the world. Hence, ecologists challenge the notions of stability, resilience, balance and cycle that characterize the phenomenology of fisheries and of the exploited ecosystems and upon which is based the initial approach of marine fishery [PAU 02, PAU 06]. Nonetheless, no acceptable approach has yet been formulated to replace the one which inspires the exploitation economy of the resources, referred to as renewable, since the 1950s. The attempt of establishing an approach based on the premise that we can adapt the conventional exploitation models of non-domesticated resources with a sophisticated set of regulations on fishing seems unrealistic.

It is thus increasingly apparent that it is the complex organization of marine ecosystems which makes the statistical estimates of stocks difficult, and the forecasts improbable [MCG 01, THE 98]. In particular, the MSE is impossible to estimate in practice because the mortality in fishing depends not only on the fishing effort deployed but also on a catchability factor which results from numerous physical, technical and behavioral parameters variable in time and space [FRE 99]. These same factors make the biomass (B) and fishing mortality (F) evaluations inaccurate, and as a matter of fact, rarely implemented. As such, the management by the target reference points based on F_{MSY} and/or B_{MSY} present the same inaccuracies. The models, too simple and too reductionist, are confronted with the insufficiencies of the concepts of exploitation, resource, inventory and sustainability that they serve. The natural complexity and unpredictability force to maintain exploitation at a very low level, in a spirit of precautionary approach. In theory, the point is to consider that the capture of a small number of fish, regardless of the technique used (industrial or small-scale), changes the ecosystem and the variability of its dynamics, making the calculation of its resiliency quite evasive. In order to adapt the ecological parameters to management, it is possible that many trials may be needed. In the best-case scenario, management will be reactive, that is to say based on the rules of short-term decisions in function of the state of the observed stocks. However, the associated uncertainty, the technical feasibility and the monitoring costs do not allow for generalizing of these approaches.

2.1.2.3. *Ecological place of humans in the marine ecosystem*

Fishery models often present fleets as an extension to the human species, considered as a super-predator of the food chains that it exploits. The fact of considering fishing as being or not part of the marine ecosystem does not seem to have been solved in fisheries research [FOL 98]. However, from a theoretical point of view, the human species cannot be reasonably considered as being part of the marine ecosystem in which it executes only a small part of its activity. We will demonstrate this in section 2.1.4. Other species are in a similar situation, notably sea birds or pinnipeds (seals), reproducing on the coasts but feeding themselves exclusively at sea. They can be regarded as terrestrial, exploiting the neighboring ecosystem, or marine, exploiting coast for social functions and reproduction. The flows of matter and energy that they generate establish an exploiter–exploited relationship between the two systems, and they are located at their interface. Eels (see Chapter 4 of [MON 14a]) present another example of species located at the interface between two systems and whose balance of energies and ecological interactions can be established. Through these examples, it can be emphasized that the ecological interactions between these two worlds are carried out by the individuals alone, with the means of their own bodily metabolism. On the contrary, the human has a special characteristic which is to use tools (prostheses) and machines (fleets) supported by a very powerful metabolism and supplied by resources essentially domesticated on the terrestrial domain. As a result of these means, the interactions with the exploited biomass are not corporal or individual but depend on the tools employed. If fisheries research wants to progress with an energy efficient approach of the provisioning of human societies with seafood products, it will be necessary that it qualifies more precisely the place of the human species in the systems in which it participates, the metabolism at stake and the associated evolutionary processes. By clarifying the theoretical and accounting frameworks of the ecological place of humans exploiting the ocean, it will be easier to make a quantitative reasoning about the conditions for the sustainability of its mass provisioning in marine resources.

2.1.2.4. *Incorrect accounting of the environmental and societal impacts*

The world demand for marine resources, and aquaculture products, is experiencing a strong growth related to the human demography and to the increase in the average standard of living. The average fish consumption has increased from 9.9 kg per capita per year in the 1960s to 18.4 kg in 2009, which represented on this last date 16.6% of all animal protein ingested by the human population [BRO 03, FAO 12]. Provisioning allows to satisfy this demand due to the globalization of the business, to an organization increasingly technical of the supply chains and of the aquaculture production of marine species, including on land. As a result of increasing energy and power supply costs, only the aquaculture of species with very high added-value justifies these facilities, or the exploitation at sea with extensive methods that do not modify too much marine ecosystems (bluefin tuna in Australia). As the other provisioning supply chains of industrialized human societies, that of seafood products and of aquaculture sees its sustainability being challenged by several findings:

– the quantities of fossil energy necessary for the functioning of this sector are very high: the energy efficiency of the chain is not good [PAU 02, PEL 07, SUM 08, TYE 05], but in fact further aggravates (increase in fishing effort due to the scarcity of the resource), and the fleets are in overcapacity [FRE 08, PAU 02];

– the impacts on the environment are significant not only through the release of greenhouse gases, but also through the degradation of seabeds (trawls effect) and, at *pro rata*, due to the activity of fleets and of ports (coastal facilities, toxicity of anti-fouling paints);

– the ecological impacts of overfishing are starting to be taken seriously into account (drastic reduction of biodiversity, proliferation of harmful species and decrease in productivity) due to the fact that they coincide with overfishing and the scarcity of catches;

– the income distribution of the supply chains is done to the detriment of the actors at the beginning of the chain when the organization of the market is not local and of small-scale [VAL 07];

– the purpose of fish exploitation for human development is being questioned given the diversion of local resources toward insufficiently remunerative export supply chains, the effects of contamination (dioxin, PCB, etc.) on health and the accentuation of the issues of food security in very poor countries [FAO 06, FAO 12].

2.1.2.5. *Management challenges of the supply chains going beyond the framework of the initial approach*

Despite this severe questioning, the initial model of the fisheries is firmly rooted, notably during the negotiations conducted by the regulating states that attempt to arbitrate the views and interests of the actors involved. Often, the fishing policies tend to "kill the messenger", such as the scientists or the experts estimating the resource, so as to ignore the fact that only radical solutions can have an effect on the reconstitution of stocks, especially in the long-term. Often, scientists are assigned an impossible role which consists of fixing the parameters of an equation which has little meaning on the biological plan. Their recommendations are then rejected or adapted, depending on the degree of authority of the political power. The social and political pressures are strong enough to sustain or save the professions at the upstream of the chain, or even the survival of poor populations which do not have other alternatives to survive. The market's economic pressures are just as considerable due to the worldwide provisioning demand of rich and emerging countries, at the end of the chain. This is without mentioning the role of sea and of littoral users outside the fishery sector. The policies are finally directed by local and sectorial compromises, without overall effectiveness, deformed by distortions and the legal, economic or geographical contradictions which challenge is never sufficient to be reformed [MES 08].

The processing supply chains of fishing products are rather opaque and it is often difficult to reconstruct the factors of price setting, the effectiveness of the transformation processes and the losses. However, studies do exist which define the mechanisms involved in a few emblematic supply chains [FAO 06, GUI 08]. Currently, the reviews effected on a worldwide scale are above all the ones which reveal the challenges of the supply chains reorganization by producing overall (aggregated) and plausible estimates on the value chains of the world market [GUI 08, TVE 05, TVE 06, VAL 07].

The economic valuation of fishery resources in the value chains located downstream are in part not included in the fishery economic analysis. As in other agri-food sectors, the prices are not reflected in a balanced way along the chain. The upstream actors (fishers) are struggling to create added value while the actors located downstream appropriate the bulk of the wealth created [GUI 08, RII 08, TIE 05, VAL 07]. However, as in most of the agri-food industry in a globalized liberal economy, the provisioning supply chains of wild fish are in crisis [FAO 12], despite a growing demand. With the development of export markets, globalized and copying the standardized markets of raw materials (commodities), the marine fishery activities can no longer reason regionally, according to a production logic centered on fishing and its landings in the harbor.

One of the industrial applications of fishing is the milling industry, for the production of fish meals and fish oils destined to the sectors of meat production, of continental intensive aquaculture and to the whole world agri-food supply chain (see Chapter 4 of [MON 14a]). This market is accessible to fisheries which are capable of large regular volumes and which accept the introduction of financial tools and of a market quotation on the stock exchange. These markets are favorable to a sector that can be criticized regarding the costs in fossil energy (cost in energy per unloaded ton, processing in fish meal plants, transport, storage), the effects on climate change (carbon balance), and the poor energy efficiency of the final meat or farmed fish (proteins) product [BRO 03, PIM 84, RAM 02]. In addition, the added value remains relatively low, in a context where the worldwide consumption of meat generates diseases implying expensive cares [BRO 03]. The environmental and overall economic balance of these fisheries has only been done in few countries (see the example of the Peruvian anchovies in section 2.2.2). The stocks of small pelagic species are the ones that most often display the largest interdecadal fluctuations already mentioned, regularly raising concerns about the stocks collapse (case of the Pacific sardine in the 1960s) [FRE 08, RAM 02]. The mechanisms that direct the phenomenology of this type of fishing and its supply chains are complex and, in the case of Peru, fairly well regulated despite the overcapacity of the fleets and the plants associated with this activity. Marine fishery research thus can no longer focus on the fishing or aquaculture activity according to the traditional logic of production mentioned above when the logics of sustainable consumption are developed which invert some reasonings about organization based on the consumer's demand.

2.1.3. *Contributions of the new sciences on sustainability and responsibility*

For the purpose of seeking solutions to overfishing, it would be necessary to give more substance to the calculations of the physical, biological and economic properties for the entire provisioning process of human societies, based on land, which are not therefore part of the ocean system. However, scientific ecology has difficulties in modeling these relationships because "naturalist" biologists-ecologists are not familiar with the biophysical properties of the dominant artificial, industrial and technological system. Conversely, specialists in industrial production and in civil and environmental engineering have only an elementary idea of the ecology of natural ecosystems dating from the works of the Odum brothers, in the 1950–1970s, on which we will come back [ODU 71, ODU 07]. Since then, scientific progress has been made to understand the complexity of nature and to try to model, according to similar principles, an industrial and domestic metabolism of the globalized, urban, economic, technological, growing and developing human society (within the biological or ecological meaning).

We will demonstrate the practical (accounting) interest in considering fishery activities and aquacultures as part of the Earth's system of human societies which exploit the ocean system without being able to domesticate it. But, the fishery case also lends itself to an ecology theoretical demonstration. For this, we will rely on the contributions of the most recent developments and progresses of classic naturalist ecology, industrial ecology, ecological economics (EE), the new thermodynamics ecology and a GE to be confirmed as a synthesis of the previous ones (Table 2.1). The integrated evaluations of sustainability attempt to define an overall performance inspired by the environmental management and the eco-social approaches of sustainable development. These last two concepts have generated a new management science field: the management of the transition to sustainability and the responsibility that we will mention for the construction of scenarios.

2.1.3.1. *Industrial ecology contributions*

IE originally proposed to describe the characteristics of material and energy flows that characterize the industry and its artificial production [AYR 78, AYR 94c, SOC 94, WEI 09]. Two major concerns motivated the first studies of material and energy flows: the control of pollution sources by the administrations, particularly in the U.S., and the rationalization of production systems, to control the costs and the pollution. Quickly, the question of industrial and domestic waste has raised the concerns of recycling (closure of matter flows), of dematerialization and of energetic efficiency. These calculations have reinforced the founding intuition that the industry could imitate nature and form a symbiosis with natural ecosystems (biosphere), or at least reduce its footprint on the resources of the planet for a same socio-economic benefit. The theoretical bases of IE were not very strong when they have been developed by engineers [GRE 10, WAC 96], but the metaphor of the IE was born. Since the 1980s, modern IE [ADR 97, ALL 99, AYR 02, SOC 94] has nonetheless become an academic field in itself, notably by strengthening its thermodynamic bases [AYR 94a, AYR 98, AYR 02, BAC 10, ORO 96]. It has created international scientific journals including the *Journal of Industrial Ecology* and academic societies including the *International Society of Industrial Ecology* (Yale, USA). Today, IE has expanded with major contributions which cause its concepts to converge with the eco-energetics of naturalist ecology [DUV 80, JOR 00, KOO 10, ODU 07]. Curiously, because of the effects of the growing specialization of research, synthesis has not yet been formulated, notably because the energetic balances of natural and industrial systems do not follow the same structures. We will subsequently develop this demonstration point.

IE generally ignores the energies of the bodily metabolism of humans, focusing on the most quantitatively important energy flows, except if human physical work plays a direct role in production (in the same way as draft animals and machines). In natural systems, few species resort massively to domesticated energies or even to processed materials (chemistry, industry and technologies). On the contrary, these means directed or driven by human thought appear to complement the rather limited metabolic energy of the human species whose evolution has always relied on tools and other exosomatic prostheses. However, unlike other animals, these human resources are not produced with the primary and secondary energy generated by food (assimilation and food chains). Depending on the authors considered, IE is a dead end when it comes to the corporal energies of the human species, nor does measure their piloting functions which relate rather more to the embodied information of the system than to its energy: IE cannot become the science of eco-energy synthesis.

Having no access to the industry data, the specialists of natural ecosystems have never been able to study in detail artificial human systems, in spite of precocious attempts to which we will refer again later. It was not until the 1990s, that Robert Ayres [AYR 94a] proposed another metaphor, "the industrial metabolism", in an attempt to reconcile the energy-efficient or biophysical calculations of naturalists from those of industrialists and economists. Today, the metabolism calculations are carried out on the whole system of human societies, at regional or worldwide scales, and define an "industrial and domestic metabolism" [BAC 10, GIA 98, GIA 07]. But, each discipline has its own conventions with regards to energy balance which prevents the synthesis, despite the proliferation of unified or GE theories [FRO 08, JOR 00, JOR 07]. We will come back to the convergences to expect from the different currents of modern ecology.

IE has generated research on the impacts of polluting emissions and waste on the environment and human health which are converging with conventional naturalist ecotoxicology. The methods of the first studies of the impact of pollution on the environment are based on the "material and energy flow analysis", or MEFA, which have generated several families of methods (mass-energy balance). The MEFA do not deduct the impacts on the environment, but remain the basic tool of the evaluation of the industrial, then of the domestic metabolism, or of the biophysical properties of economic systems. Therefore, the MEFA not only allows to calculate at the scale of production units, but also of regions or states, the "Total Material Requirement" (or TMR) which corresponds to the real material flows in the economy, expressed as mass (tons). Accounting conventions (input–output*)* thus allow considering the inputs of mineral resources, biomass, water and land provided by ecosystems, and the outputs in subproducts, waste, emissions and discharges, with regards to economic performance [ADR 97]. Von Weisäcker in Germany and Lovins in the United States [WEI 09] were pioneers of these analyses. They have very quickly produced an assessment and pragmatic concepts of environmental management to respond to the issues of the use of limited resources, originally raised by Meadows *et al.* [MEA 72] for the Club of Rome in the 1970s (Club of Rome, 1973), and then by the concepts of the "4, 5 and 10 factor" of which they are also the authors.

Lifecycle analysis (LCA) constitutes the central method for combining the inventory of materials and energy flows in the industry and the quantitative assessment of the associated impacts on the environment. The latter are estimated potential impacts (non-directly measured) and they are related to a mass, service or functionality unit. Categories of impacts have been defined with reference to the broad worldwide mechanisms of the degradation of the atmospheric layers, of the air quality, of the climate, of soils and their chemical properties (acidification, contamination, etc.), and even of the degradation of the major ecosystems and the biodiversity, or still of the depletion of abiotic resources. Through LCAs, it is possible to compare products, processes, industries or services through their functionality (functional units), whose explicit definition obliges to verify that the comparison has a meaning. From this point of view, the production–consumption perimeter to consider for each of the compared objects constitutes a crucial element. LCAs have allowed dreaming about the possibility of endlessly recycling in order to drastically reduce the consumption of raw materials, the production of waste and to try to save the industrial model through a reorganization of the industry and of "mass" consumption [AYR 94b, AYR 02, BRO 03, WEI 09]. LCAs also enabled the demonstration of the limits and advantages of some promising concepts such as recycling, remanufacturing, clean technologies, the partial dematerialization of the economy, energetic intensity, the use of technology, the increase in the share of services on objects or the substitution of inefficient products. For example, it is shown that in most cases, the sustainable or the "ecological" properties of products and services can only be defined according to the environment or to the system in which they are used (from the cradle to the grave). Through these concepts, it is possible to act and to guide public policies or the coherence of actions in enterprises and at the end user level. We refer the reader to the synthesis proposed by Robert [ROB 02] on those applications which, to be operating, must be embedded into an economic and political process (policy-making), and then eventually into a participatory or concerted process of operational project management.

The purpose of the multi-criteria methods of impacts assessment is above all to translate the considerable mass of scientific and engineering knowledge into indicators, descriptors and synthetic or agglomerated indices. Such accounting tools sometimes stray away from scientific methods that prefer deterministic quantitative IAM-based models (Integrated Assessment Models). The simplifications follow explicit conventions and allow the translation of expert knowledge to inspire engineering, management modes and policies. These simplifications justify the sacrifice of part of the precision and the theoretical rigor, for example through the creation of dimensionless indices from heterogeneous quantitative data. These same simplifications and the lack of up-to-date data make these methods questionable [AYR 95]. The methods of the environmental impact assessment, or EIA, have taken on, since the 1990s, the challenge to gather data from the industry (production) and from the global consumption into dashboards or databases (for example EcoInvent). Multi-criteria analysis software can exploit them (e.g. Simapro, Gabi and Umberto) [PRE 08] in order to produce indicators that can be freely chosen according to the political challenges [FAU 98] or to sectorial specificities.

The approaches known as "integrated" propose today to conduct evaluations both of the environmental criteria and of the social criteria that drive economics, which parameters are also extracted. The Lifecycle Sustainability Assessment (LCSA) method developed by the *Society of Environmental Toxicology and Chemistry* (SETAC) in the USA is consensual, in institutions as well as in the industry (see the website www.unep.org), to build systems of aggregated indicators with explicit accounting rules and resulting in the composite indices of the overall performance. We will come back to this in section 2.2.

2.1.3.2. *Ecological economics*

EE is born out of initial theoretical works from Harold Hotelling and other thinkers of the 1930s, and then from Nicolas Georgescu-Roegen in the 1970s [GEO 71]. EE has merged a set of converging currents from the economic theories of the environment and natural resources [FAU 97, TIE 00], and then from a criticism of the dominant, neoclassical and liberal economic thought [PAS 79, PAS 10]. These theories had inspired the juridical profusion of the 1970–1980s about the environment in industrial countries, but they were initially not challenging the neoclassical view of economics. They simply insisted on the necessity of taking into account the risks and the damage on the environment and on human health in the economic calculations, but also of considering that the updating of the environmental preservation costs justifies taking a few risks for the production and to postpone its invoice to an unknown date. These are thinkers, such as H. Daly [DAL 11] and more recently Costanza [COS 02, COS 03], who have gradually brought EE out of the neoclassical version of economics, not without causing controversy that tended to normalize after the generalized crisis of 2007 [GER 12]. Some authors demonstrate that the transition from an economy of production toward a responsible economy of consumption can only be achieved if the driving forces of economic growth integrate the productivity of the resources and the responsibility of the industrial actors with regards to what the products they design might become [AYR 10, EPA 06, EUR 10]. Then, the economics would include the incentive forces sufficient to generalize the ecoconception and the eco-restructuration of procurement and consumption supply chains of human societies. This involves taking into account the regional inequalities of access to resources, the environmental sensitivity and the development objectives.

Many currents of thought, qualified as heterodox, have emerged, including that of bioeconomics supported in France by René Passet [PAS 79]. EE is still immature in its theoretical foundations, even if it has generated an already considerable quantity of scientific publications, particularly after the major financial and economic crisis of 2007. EE attempts to develop the tools of economic analysis to develop the neoclassical economic or dominant liberal thought about the issues of the natural capital, of energy efficiency, of sober and prosperous growth and development factors, of the externalities and the hidden costs of capitalism. This list also includes the issues of biophysical limits of growth (production), of the fairness of wealth distribution, of the services achieved by the natural ecosystems, of the productivity of resources, of the sustainability criteria of the activities or even of the governance modes [BAR 08, BER 04, FIS 07, GOW 94]. These subjects have become those of most of the reforms of capitalism, launched implicitly by states, after the financial crisis of 2007. EE is however not recognized as the theoretical framework of this silent revolution [COS 02, GER 12].

There is a convergence between EE and IE on the operational concepts of the management of sustainable development, generally empirical. These concepts engage the economic system in its physical and biological foundations through the flows that it generates, and must inspire the transition strategies toward the sustainability of human activities. They bring insight about the possible articulations between ecology, economics and IE. The strength of EE [BER 04] has consisted of gradually validating, in a pragmatic way, the experiments conducted for decades by a few pioneers who had not theorized their results into renewable energy systems, biological agriculture, the extensive uses of space, an efficient domestication of natural resources (see section 2.1.3.4) [KRO 07].

2.1.3.3. *Toward a generalized ecology (GE) of the biosphere*

What is called the *new ecology* has been proposed in the 2000s, by specialists in eco-energetical balances and in modeling the complexity of biological systems, gathered by S.E. Jorgensen [JOR 07, JOR 09, JOR 10]. These authors have focused on the thermodynamic energetic approach and on the very general properties of ecosystems, neglecting to detail the consequences of their contributions on the classical concepts of naturalist ecology. The Odum brothers had, since the 1950s, attempted to generalize the basic concepts of cybernetics, of the thermodynamics of dissipative systems into eco-energetic approach supposed to authorize the easy energies balances of natural and human ecosystems. But their representations are not rigorous with regards to the principles of the thermodynamics of open systems, poorly identified at this time. Nevertheless, they have profoundly affected scientific, then political ecology, around the concept of ecosystem. Their primary approach [FOL 98, ODU 71] has proved to be not sufficiently strong with respect to the complex properties of living systems, discovered later. A pathway is still necessary to fully integrate the recent developments on the complex properties of systems [NIC 77, SCH 94], although in the 1920s, Verdnasky has defined the biosphere in very visionary terms to characterize the model of the planet ecology, often called global [SMI 08, JOR 10]. He had understood the energetic challenges that are used as driving forces in the life process and its complexity through the evolution of species and ecosystems, according to a continuity of scales. Downscaling below the biosphere level (ecosystems) seems, as a matter of fact, quite artificial and makes theoretical, practical (engineering) and political (human ecology) ecologies almost contradictory because of the many accounting conventions of ecology currents. The literature which has enabled the beginning of synthesis in the 1990–2000s is very abundant, but remains fairly compartmentalized in historical disciplines. The unification of the ecological sciences has difficulties replacing the conventional schemes, even though they are too elementary [FOL 98, ODU 71, ODU 04]. Eugene, and then Howard Odum, proposed a framework for the unification by defining the *Life Support Systems* [ODU 07], but the adopted thermodynamic and systemic simplifications are simply unacceptable by todays standards. Following an ecological planning approach, Paul Duvigneaud [DUV 80] proposed to define the "anthropized" systems of "Urbs" (urban), "Ager" (intensive crops) and "Saltus" (prairies and livestock) as "ecosystems directed and modified by man", but the approach was empirical and has not asserted the emergence of these new systems in the ecological theory.

A French author, Serge Frontier [FRO 99, FRO 00, FRO 08], has significantly contributed, under the influence of the Spanish marine biologist Margalef [JOR 10, MAR 68, MAR 05], to clarify the classification of types of energy and matter flows that characterize both the natural ecosystems and the artificial humans systems.

Frontier summarizes well these fields of reformulation of the theory to "ten keys to the analysis of ecosystems" [FRO 99][3]. They are:

– the interactions between species and with their environment;

– the diversity of strategies of these species;

– the flows and cycles of matter and energy;

– the various nonlinear and unpredictable dynamics and their effects;

– the internal (embodied) information contained in the biodiversity which is then regarded as a fractal dimension of biomass;

– the space-time fractal occupation;

– the hierarchical structure that reveals the importance of connectivities (connectedness);

– the relationships between the physical environment and the biology of communities of species, and their influence on the transfers and the processing of energy and biomass;

– the evolution of ecosystems and communities;

– the interactions between ecosystems in the biosphere.

Two major contributions of this eco-energetic reformulation work are:

– the definition of auxiliary energies (and of the auxiliary biomass that we will not cover here) that does not go through the primary and secondary production of food chains and of organisms metabolisms;

– the exploitation principle of an ecosystem by another, which places them on the same hierarchical level in a larger ecocomplex such as the biosphere.

The concept of "auxiliary energy" is derived from the naturalist eco-energetics in its recent version [FRO 08, JOR 10, MAR 05]. It meets the concept of natural or artificial forcing of biological

3 Developed in later publications, see [FRO 00] and [FRO 08] (supporting documents downloadable from Dunod's site: www.dunod.com).

processes, as used by the specialists of major biogeochemical flows in global ecology [SMI 08]. Ecosystems are structured by the energy assimilated or metabolized by the organisms along food chains in primary production (autotrophs dependent on photosynthesis) and secondary production (heterotrophs, predators or scavengers) in biomass. However, the organisms also use, directly or indirectly, inassimilable energies which do not participate directly in the life process, but accompany it (energies of "covariance" with the environment, the covariance reflecting the spatio-temporal coincidence of the environment and of organisms' variations). The "primary auxiliary energies" are provided by the physical environment (climate, currents, materials and environment chemistry, etc.) that the organisms select as its living environment, with the evolution of species. As a result, organisms can also dedicate a part of their endosomatic metabolic energy to means that better use the nvironment and gain an individual or collective advantage for their metabolism. This energy is called the "secondary auxiliary energy" and corresponds to: structural energies used to bring together the photosynthetic resources (in a terrestrial environment only), energies of active motion for feeding or access more attractive areas with regards to food, oxygen or reproduction, and agitation energies of the environment (bioturbation and human domestication) which change the productivity of the environment, the construction of tools or of rigid domestic structures (nests, anthills and cities). Every organism therefore adopts adaptation strategies during the course of the evolution of species that take advantage of the facilities of the environment and compensate for the constraints of survival. However, this has an energy cost (or benefit) which weighs down on the metabolism, especially if the primary auxiliary energies (issued by the environment) would not be in covariance with the needs of the organism. The metabolism evolution of organisms thus depends on a dynamic compromise between a good use of assimilable energies (food chains initiated by solar radiation) and the energies of the environment, referred to as "auxiliary" or of natural forcing. The auxiliary energies are generally neglected by environmentalists in their balances, particularly because their measurement is difficult [KOO 10]. The metabolism of living organisms, their body dimensions (allometry), their strategy in food chains and in ecosystems are largely directed by these auxiliary energies, that put matter and resources into motion. The auxiliary energies are therefore the thermodynamic driving forces of the self-organized reactive process of life and represent phenomenal quantities, whereas the biomass energy flows in ecosystems (life) are extremely low [SMI 08]. In conclusion, the measurable endosomatic metabolism of living organisms, their corporal proportions, their adaptive strategies in food chains and in ecosystems are largely directed by the auxiliary energies that put the matter and resources into motion without determinism. Their contribution to the production of biomass is essentially qualitative and informational. They complicate the accounting of natural metabolism which should respect the principles of thermodynamics to produce scientifically relevant energy parameters [JOR 10, KOO 10].

The concept of auxiliary energy allows classifying the energies mobilized by man in a consistent manner with an energy balance at the scale of the biosphere (*new ecology* of [JOR 07]). The GE, partially formulated by Frontier ([FRO 08] in Chapter 9) recognizes an industrial and domestic human metabolism (human societies) consistent with what naturalist ecology proposed, in its current version. However, the theory has never been formulated explicitly to propose a generalizable energy-efficient balance. Nevertheless, GE is the most appropriate theoretical framework to relate the biophysical parameters of the overall economy (formal and informal) with those of the worldwide ecosystem (biosphere), explicitly including the industrial human species ("anthropocene", see Appendix 3) and its impacts as they are considered negative on the original ecosystem. Even more than primates, than social insects or corals, the human species is collectively capable of, as we have seen, mobilizing the "exosomatic" means that are the objects (material culture [WAR 99]), tools or prostheses, but also the substances which leverage the basic means resulting from the use of hands and of the body. The domestication techniques of materials and energies available in nature (wood, peat, fossil biomass, physical and chemical energies) are driven by thought, which is an expression of the brain, an organ that consumes 20% of the energy of the corporal metabolism.

Systems	Energies of a generalized assessment of the biosphere	Energy density (orders of magnitude on the basis of W/m^2) [SMI 08]
Exploited natural ecosystem	Species metabolism: part of the basal metabolism (development), maintenance and reproduction	++
Exploited natural ecosystem	Secondary auxiliary energies of species (share of the mobility and the structures metabolism)	+
Exploiting human ecosystem	Human corporal metabolism (reported in human societies)	++
Exploiting human ecosystem	(Including driving energies or part of the corporal metabolism of machines control)	(ε)
Exploiting human ecosystem	Human secondary auxiliary energies (especially exosomatic) useful for procurement (low efficiency)	++++++
Biosphere	Primary auxiliary or natural forcing energies	+++++++++++++++
Biosphere	Dissipated human secondary auxiliary energies of anthropogenic forcing	+++++++++++

Table 2.2. *Presentation of the energies considered in the human natural and artificial metabolism, and of the orders of magnitudes indicative of their relative weights*

The energies and materials collectively domesticated are placed at the disposal of individual human beings by a partially globalized economy. They increase in energy equivalent and exceed by several orders of magnitude that of the corporal metabolism (Table 2.2 and Appendix 2). All this energy is akin to the secondary animal auxiliary energy and is qualified by Frontier as "human auxiliary energy" and meets Ayres's industrial metabolism concept [AYR 94b] and more recently Giampietro [GIA 98] concept of IE. It allows humans to domesticate and gather the resources, and particularly the biomass, of exploited and therefore modified systems (agriculture, oceans, etc.) to feed as an heterotroph or "predator" (secondary production). In doing so, the *Homo sapiens* distorts the sets of energy networks of the ecosystems in which he participates. These calculations must inspire another interpretation of the biological productivity (biomass distribution) and of the meaning of the biodiversity in the systems modified by man. He is an emergent species that acts through a modest corporal metabolism, as a driving force on very powerful and globalized systems of machines, where it is observed that they can compete with geological or climatic forces and even change their course, through artificial forcing [SMI 08]. As a result, human "artificial or mixed ecosystems" acquire specific energy characteristics which differentiate them from the natural ecosystems because of inventions and human intentions. The definition of an industrial and domestic metabolism of human societies includes mainly these secondary auxiliary energies mobilized by human know-how. Such an approach implicitly imposes a new level of organization among worldwide ecosystems: the system of human societies.

GE has stressed the importance of the exploitation of an ecosystem by another to facilitate the description of the interactions between ecosystems, from the most local to the global scale, while the intermediate sectionings are often unsatisfactory. Recognizing the systems at the same hierarchical level is an important legacy of naturalist ecology, then of sciences of complexity. This representation allows characterizing the relationship between the emergent human species and nature, within the biosphere (Appendix 3). The theory proposes to consider that two ecosystems can exploit one another and evolve differently, since they are not in the same state of maturity or *a fortiori*, that they do not have the same biotopes and biocoenoses. Some species take advantage of distributing their activity on several entities or systems that structure the biosphere. The exploited system undergoes a disturbance more or less important which tends to rejuvenate it and to maintain it growing (rejuvenation state having an impact on the strategies of species and of biocoenoses). The exploiting system takes advantage of substantial biomass that feeds its own growth to acquire a maturity and new structures (internal information), and to accelerate its development. The distinction between growth and development corresponds to an energy allocation in, respectively, the quantitative (biomass, productivity) or qualitative (species, structures and interactions) characteristics [FRO 08, ULA 00]. Holling's general theory of adaptive systems [GUN 02, HOL 86] proposes an essential notion:

panarchical systems. This theory (panarchy), generalized to all systems, human, natural or artificial, puts forward the principle that the dynamics of a system can be followed according to three axes: potential (quantities), connectivity (diversity of elements) and resilience (capacity to withstand disturbances). A system therefore fluctuates around growth (exploitation), maturity (conservation), release and reorganization stages at multiple temporal and spatial scales according to a loop which oscillates on the three axes. Growth and development determine the evolution of ecosystems and the constitution of eco-complexes, i.e. of ecosystems nested inside one another, or next to each other, according to a dynamic hierarchical organization, made of destructions and reconstitutions of organizations from the species of the fauna and flora and microorganisms regionally available [GUN 02, HOL 86].

The flow of biomass and energy from one system to the other makes structures possible that the exploiting system could not support without this additional assimilable contribution of energy. The development stabilizes the growth necessities of the exploiting system to a level which is sufficient to maintain the acquired structure and ripens the interaction between the two systems. The exploitation mode becomes more complex, due to the transfer of information or of structures toward the exploited system. The exploiting system can be dependent on the exploited system or exploit several systems alternately. The exploitation maintains a differentiation of the two systems that can stabilize in time and space, according to a regional meta-balance (regional ecocomplex). The two systems are partners and the interface organizes the transfer of matter (biomass) in one direction and the structuring information transfer in the other, by lengthening the trophic chains and the networks. The complexity of the whole is held by a co-evolution of systems elevated at the same organization level. It can lead to a symbiosis or a simple cohabitation in a higher hierarchical level system (ecocomplex).

This reformulation of the energies at stake in ecosystems solves the shortfalls of naturalist ecology, the abstractions of the new thermodynamics ecology and the approximations of IE. GE can unify naturalist ecology and IE with more rigorous eco-energetics capable of integrating the metaphor of the human industrial and domestic metabolism. We believe that we have achieved the demonstration of the points of interest of the auxiliary energies and of the recognition of the hierarchical level of the human societies system to set up the generalized eco-energetics of natural and anthropogenic ecosystems which constitute the biosphere (Table 2.2). GE can also trace the path of a wider unification, as formulated by the sociologist Edgar Morin [MOR 80] and which would explain the biophysical bases of human ecology or the eco-social, or simply the anthropological, approaches of economy and politics [BAR 08, BER 04, FIS 07, GER 12].

The energy-efficient calculations which recognize the auxiliary energies or their effects upon metabolisms and the emergence of systems of higher hierarchical level in panarchy, allow combining the partial theories of the new ecology and GE (strongly affected by naturalist inputs) with those of IE and EE. This unification of the ecological sciences can then describe with the same energy and systemic (metabolic) terms, the relationships between the system of human societies (exploiting) and the natural systems of the same hierarchical level, exploited under the economic provisioning. Some authors coming from very different backgrounds [GIA 07, KOO 10, ODU 07, ULA 00] propose balance methods about the metabolism of living organisms open to energetic analysis at multiple scales, from the organism to the ecosystem. The dynamic energy budget theory, or DEB, has this ambition (see Chapter 9 of [KOO 10] for a beginning of extension to the ecosystem), but the conventions seem difficult to extend to the industrial metabolism of humans ([KOO 10] for example) due to a poor understanding of the effects of auxiliary energies [BAC 10, GIA 98, GIA 07]. As a matter of fact, the metabolic calculations of biologists (allometric criteria, balance on organisms) do not assess very well the auxiliary energies, not even the secondary. The "industrial and domestic metabolism" derived from IE being, as we have seen, mainly constituted of secondary auxiliary energies (human), it cannot follow the DEB theory. The two energy-efficient approaches are therefore not compatible and do not serve the same developments of ecological theory.

2.1.3.4. *Contributions of environmental management and eco-social approaches: toward a transition management*

EE has generated a large number of empirical and experimental management concepts, to make durable development operational at the scale of enterprises, territories, countries and the world. The most famous concepts come from environmental management [ISO 14a], also called eco-management. They have been established on bases often distant from the scholarly calculations of industrial metabolism theorists, but nowadays, a socio-ecological movement, that we have already mentioned [BAR 08, FIS 07], has retrieved from the economics and the political sciences perfectly mobilizable concepts to accompany change in society. We have chosen to quote about 20 of these concepts to establish the ideas:

– the concept of impacts of human activities on the environment and more generally on the human species (health, resources, prosperity and fairness) defining sustainability and responsibility [BAR 10, COS 02, GRE 10];

– man-nature transaction: learning cycles, irreversibilities, uncertainties about the risks, vulnerability of ecosystems and of the human species, global changes [SOC 94];

– recognize the services of the ecosystems and the common goods [COS 02];

– ecological integration of human activities, ecosystem-based exploitation of resources, natural capital, and ecological approaches (generalized) of human development;

– zero-emissions principle (end-of-pipe technologies, toward clean technologies) [AYR 94b];

– recycling and remanufacturing;

– eco-conception, product life duration and optimized maintenance;

– substances substitutions, materials, technologies and non-sustainable processes;

– dematerialization, circular economy and economy of functionalities;

– production logics versus services;

– sustainable consumption and stakeholders responsibility logics;

– accountability of designers-producers on the impacts of their products or services;

– energy and matter effectiveness (energy efficiency);

– energy and matter intensity (ratio between energy and produced wealth);

– hidden costs of capitalism, societal responsibility in overall cost [ELS 06];

– decoupling between activities and unexpected rebound effects that can cancel the initial efforts of impacts reduction [BAR 08];

– failure of command/control regulations, equitable regulations or Smart regulations [GUN 02];

– governance of social and natural capital, equity, distribution of wealth, positive sum games (win-win), civilization goals;

– production (growth) necessary to the human development;

– new drivers of growth (productivity of resources, of capital and labor, abundant and clean energies, technologies role [AYR 05, AYR 10, GER 12]).

These twenty concepts have the potential to inspire several levels of reforms fundamentally challenging the question of the economic system. It is easily observed that the concepts described above are not located at the same operation levels. Some affect more the techniques to mobilize, the designs and the management of projects, others the intervention of States in the public policies and in normalization, the right and the taxation, the international agreements and eventually the constitutions [BAR 08, FIS 07]. The above 20 concepts must bring insight about the objectives and the mechanisms of the new policies because they integrate the physical and economic dimensions of activities.

The authors of the first scientific and technical IE are well aware of the difficulties to generate a change in the practices of the political organization of human societies to make their activities durable and sustainable. Some even consider that on the ground, things are not really changing [BAR 08] since the Rio Conference in 1992 [WCE 87]. The environmental management, both in enterprises, public administrations and among individuals–citizens–taxpayers–workers, is a major challenge of the action, now that our Society awareness is growing.

Today, there exists a management science which proposes management principles of the transition to sustainability (MTS) [ROB 02]. Instead of dealing with the issue of sustainability assuming that a model for sustainable economy might one day replace the elder one, the promoters of the MTS demonstrate that the economic system can change by taking change trajectories that must simply be chosen and progressively conducted.

To inspire this new management inside organizations, standardized methods have been developed from the experience of the professionals: environmental management referred to above, then sustainable development and corporate responsibility. Today, sustainable development tends to make room for the societal responsibility of organizations, assumed to encompass the whole (see [ISO 14b]).

Despite the ambitious synthesis of experts mobilized by ISO 26000 on the "societal responsibility of organizations" (SRO), the debate on the articulation of the sustainability levers (environmental and social), and of the social or societal responsibility, is not over. Sustainability establishes fairly well the technical aspect of the approach when the responsibility engages the political and the juridical attitude. The weaknesses of the responsibility attitude can be exposed when the economic activities are not insurable nor useful to society. There, lies an important field of reflection to give a meaning to the design of public policies and enterprise strategies by the better use of the evaluation methods of the sustainability of products, processes and services, inspired by the analyses of the sustainability of the lifecycle and the process of their political interpretation. We will develop this issue through our case study (section 2.2.2) and in section 2.3.

2.1.4. *New accounting framework for sustainability and responsibility: integrating uncertainty and "uncontrollability" in fishery and aquaculture management*

2.1.4.1. *Theoretical framework of fisheries exploitation in generalized ecology*

Fishery exploitation follows many forms (occasional, sportive, subsistence, small-scale, industrial and aquaculture) that contribute more or less directly to the global provisioning of the human system based on land. The survival of regional and small-scale fisheries does not escape the globalization of the human economic system. As a matter of fact, if fishing becomes the only resource for the survival of some populations, to the extent of causing overfishing, it is in part because of the failures of the world economy to offer them provisioning alternatives.

Although the spatial behavior of a fishing unit imitates that of superior marine predators (super-predator metaphor) [BER 05], we highlight three arguments based on section 2.1.3.3 to support the thesis that "the person who fishes" cannot be considered as part of the "ocean" as an exploited natural ecosystem:

– the secondary auxiliary energies who arm the fleets do not come from the ocean ecosystem;

– the exported biomass of the ocean feeds the growth and the development of the exploiting system of terrestrial human societies, but does not determine the energetic capacity of fishermen to come back fishing (and generally exceeds it by several orders of magnitude);

– the direct and indirect environmental and ecological impacts (pollution, climate change) of the terrestrial exploiting system toward the exploited ocean ecosystem correspond to flows of matter and human auxiliary (secondary) energy which change the natural forcing of the ocean ecosystem by anthropogenic forcing (also called artificial). These impacts create another type of interaction between the two systems, energetically remote and evolutionary;

– the disruptions of the exploited system by the exploiting system are not discriminating (types of interactions between food chains species, biodiversity adjustment). They rejuvenate the ocean ecosystem (in the panarchic ecological sense) and make it evolve in such a relatively unilateral way as long as the fishing pressure does not diminish. If it does diminish, a new adaptive learning system of the exploiting system begins.

A harmonized eco-energetics approach allows to reason in a GE, which is very rich within the nature of the interactions between human societies and the ocean ecosystem. Frontier [FRO 08] is very explicit on this interpretation of the ecological systems theory. The relation between two ecosystems, or eco-complexes, no longer has the same weight. That is to say, in the case when one massively exploits the other, it creates another level of connectedness where individual interactions between organisms occur. Fisheries are a very good example of the nature of the change of the hierarchical level of ecological relationships of an emergent system (the system of human societies) which develops with the rest of the biosphere as a whole.

The representation of fishery exploitation by two differentiated systems with the same hierarchical level challenges some of the deeply rooted principles of the economy of renewable natural resources, of IE, of EE and of course of fishery, whose principles of "ecosystem-based fisheries management" are very recent (EBFM; see Chapter 3 and Chapter 4 of [MON 14a]), [FAO 99, SOT 08]). Yet this representation is a justified framework necessary to integrate the recent contributions of GE as we define it. It may be considered, in theory, that any removal of marine biomass by humans with exosomatic means modifies the exploited ecosystem, and thereby introduces a new interactions cycle (dynamic and adaptive on several scales) between two systems of the same hierarchical level: human societies and world oceans.

The previous considerations lead us to assume hypotheses where the exploited system has been profoundly altered and where the exploiting system must find its form of development in the ecological sense. It could be that the two systems can one day be connected enough to naturally develop form and strategies of co-viability

[LEF 99] or even to reach an equilibrium as proposed by [MAR 07], without necessarily mentioning symbiosis. But as we have seen humans use significant exosomatic means, placing the ships, its fishing gear and the sailors outside the ecological system that they exploit. As a matter of fact, except in the case of dissuasive depletion of resources, little "feedback from experience" acts on the behavior of the fishers and, in all cases, it is not of the same nature as the selection mechanisms operating through the food chains in the ecosystem. If there is feedback from experience, translated by a few regulations, it focuses mainly on the fishing gear and on the fleets, and relatively little on the physical abilities of the sailors and on their corporal metabolism (without its secondary auxiliary energy).

An evolutionist perspective is necessary to consider the scenarios of the long-term exploitation of the ocean, taking into account the disturbance of the adaptive feature of the ocean ecosystem by the intervention of humans with very powerful artificial means. From this point of view, the general theory of Holling's adaptive or panarchic systems [GUN 02] can help to redefine the adaptive frameworks (evolutionary) of fisheries management by recognizing the uncertainty in the resource (exploited natural ecological systems) and the uncontrollability of fisheries, as an exploiting, emerging, young and learning human system. In the panarchic cycle, the coupled ocean–human societies seem to be located at the beginning of a reorganization in creative destruction. Indeed, Holling's panarchic theory offers a framework for new adaptive fishery models, which would learn from the lessons of overfishing ([GUN 02], Box 7.1, p. 175). Overfishing would then be considered, without judgment, as the opportunity to reorganize the relationship between fisheries–fish, or more precisely, the relationship between oceans–human societies, because the collapse of the resource must be brought closer to the reactions of human societies on a longer-term, in order to close an adaptive cycle. Ecological science does not presume that the new system, succeeding the collapse of the first, is not capable of a reorganization in the continuity of the life process. The irreversible degradations of the current marine ecosystems would appear relative, and would not prevent a few regrets, in particular if loss in biodiversity occurs.

2.1.4.2. *Accounting framework of the supply chains sustainability*

The requirements for fishery and aquacultural sustainability involve a precise limitation within the exploiting human system at its interface with the exploited system [WOR 09]. To achieve this, it is necessary to consider the supply chains that collect, sell and consume (from the ocean to the plate) the captured biomass which is then take ashore. However, it is also necessary to recognize those who maintain the technical and social operation of these chains, beginning with the fleets and their gear. The ideal is to strive toward a unified and central model able to deal with the direct or indirect, short and long supply chains, probably open to non-food usages of the extracted biomass. With this in mind, we can list the following limitations, which are characterized by costs and flows of energy and materials, biophysical yields, environmental impacts and contributions to economy and to human development:

– fleets and gears (construction, maintenance, and activity);

– ports and maritime logistic installations;

– trading points, auctions, markets, market stalls;

– facilities for the primary processing, the conservation (eventually embarked) and the pre-storage or packaging (at the upstream of the distribution) ;

– facilities for the non-food processing and treatment/valuation of waste;

– indirect agri-food meat or aquaculture production chains along with fish meal or fish oils but also oils with a vegetable origin, including intercontinental transport operations;

– facilities for secondary transformations (additions, cooked dishes, etc.);

– wholesale distributors (transport and storage) of all products categories;

– retail distributors (all categories not only fresh or frozen products);

– purchase operation by the final consumer (supply, deliveries and other logistical);

– storage and domestic processing;

– preparation for consumption (food ration, meals and leftovers-waste);

– human and animal assimilation (individual metabolism, reservations and compensations, respiration and contaminations);

– maintenance and reproduction functions, and activities of the human body (allowing to mobilize the human auxiliary driving energy);

– excreta and surplus discards, domestic wastes (to the *pro rata*);

– internal information of the human societies system (development and culture).

This provisioning and consumption chain is brought by an industrial metabolism of which a global quantitative review can be made (regional, national or global depending on the length of the chains). Starting from the preparation of the ration and ingestion, the chain follows the rules of the biological metabolism of the human consumer. However, food preparation and waste justify the consideration of a *household metabolism* [GIA 98], distinct from the industrial metabolism of the logistics chain of provisioning for the extraction of biomass and its transformation. This chain is the reference framework for the evaluation of the fisheries and aquaculture metabolism. This framework allows reporting the quantitative parameters of flow (materials and energy) and of impacts (environmental, ecological and social) on any unit in the chain. For example, the landed ton[4] of fishery products will be able to be compared to the ton of fresh aquaculture products, in both cases brought to the consumer's plate. The industrial metabolism at stake in procurement can be put in relation with the industrial metabolism which produces the means of fishing. The performance at the upstream end of the chain (efficiency of the fishery per kg of fish unloaded) can be evaluated [ALD 08] and, at the downstream end, a consumption strategy which would determine the fishing effort (energy cost accumulated in fuel and carbon balance,

4 Here and in the rest of this chapter, ton refers to metric ton (1000 kg).

cost of protein supply, contaminations and health overall cost). This is the approach which consists of calculating, going up along the supply chains, which food diet is the least expensive in fossil energies, the least contaminant (organic or eco-labels) and the most effective in terms of health [CAR 03] (and example of section 2.2.2).

These approaches may seem out of reach, but we have seen in the previous sections that IE offers tools for assessing these flows and proposes balances. It is in this direction that are positioned the studies of energy consumption and of fleets equipment of the FAO or of some research centers [BRO 03, FAO 12, OEC 07b, PEL 07, TIE 05, TVE 04, TYE 05]. However, the relevance of the accounting frameworks of these analyses, still uncommon, must be improved and extended to the rest of the supply chains.

The economic and organizational analyses of the sustainable supply chains can take two forms of complementary logic. The production logic justifies descending forms of balance (upstream toward downstream) with fishing means, landings and consumable or consumed biomass as its areas of work. Then, an effective consumption yield or a loss and auxiliary costs balance toward the consumer will be calculated (accumulated energy costs, impacts on the environment and on the biodiversity). The consumption or the development logics are based on ascending balances (downstream toward upstream) and rely on criteria of quality, food (proteins and food supplements), health, public policy costs, poverty, food independence[5], human development and finally of responsible and fair consumption to connect and relate to fisheries activities. In the spirit of ecological economy and of fishery bioeconomics, it is necessary to define the new forms of management of value chains in the supply chains, in the context of dual crisis of the decreasing fisheries yields and of the difficulties of the actors at the upstream end of the chain to capture a sufficient share of added value [NEW 04, RII 08, TVE 06, VAL 07].

---

5 The problems of poverty and food independence are central in countries where the marine resource represents the bulk of protein intake, but which dedicate to exportation an usually significant share of the catches, or allow fishing rights to foreign fleets in an inefficient way for the economy of the country.

2.1.4.3. *Integration of uncertainty and uncontrollability in management*

We have shown that fisheries are at the interface of the two systems with an exploiting–exploited relationship. This allows us to classify the regulation regimes relatively to two parameters of this interface: (1) the uncertainty in the estimate of the exploitable resource and (2) the controllability of the means and of the behavior of the fleets (Figure 2.1).

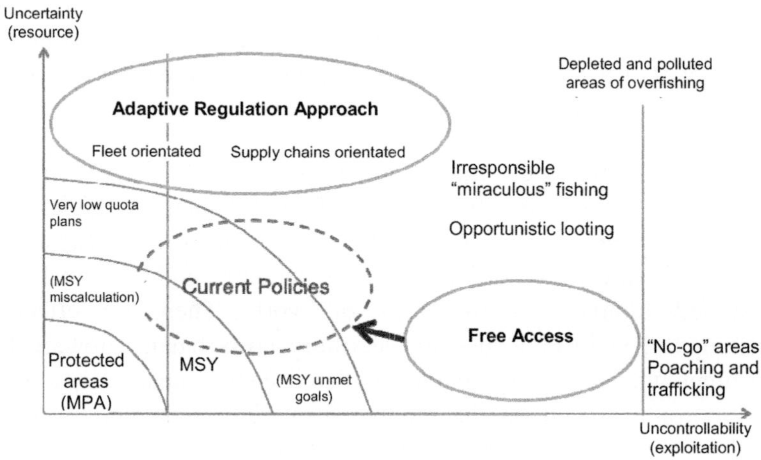

Figure 2.1. *Diagram of the management principles of renewable resources in common good*

The marine prohibited or protected areas (MPAs) are placed to the left, along the axis of the controllability and for the lowest uncertainty values (lowest part of the graph), assuming that the means of surveillance are efficient. To the right, are placed so-called no-go areas of "trafficking" activities (poaching) and their corresponding supply chains (corrupted and unofficial markets, social dumping), assuming that their repression does not work. If the latter harbor a rather non-fluctuating resource, it will encourage traffickers, because of their reaction times, the low effort to invest, and the small level of risks incurred. The zones of "miraculous" fishing, or opportunistic looting, correspond to difficult or neglected controlling conditions at the same time as a high variability of the resource (or a scarcity leading to very attractive prices). Three forms of fisheries are located between the prohibition areas and those of no-go areas, and represent the bulk of worldwide fisheries:

– free access or equivalent, whose exploitation leads to overfishing, especially if the resource has a predictable abundance in appearance (except brutal collapse);

– the areas regulated around the principle of MSY of which it is in some cases found that it was built on the inappropriate monitoring of stocks with regards to uncertainty, despite a correct controllability of the fleets;

– fisheries that achieve to apply a principle of quotas or of prohibition (MPAs) efficient when controllability is high and when allowed catches are calculated well below the estimated MSY.

Most of the current policies are lacking in certainty for the true level of controllability and are especially uncertain about the monitoring of the resource abundance. The future of fishery exploitation seems to lean toward the exploitation regimes which assume the hypotheses of the large variability of stocks and of their dependence on global properties of the ocean ecosystem (climate and chemical change, migration of species and collapse phenomena resulting from food chains disturbance). These are reactive or adaptive policies which suppose an intensive follow-up. The modes of adaptive regulation can manage the catches (quotas and control of embarked means of processing) or the spatio-temporal modes and operation modes of fleets (strong controllability). Let us notice in the latter case the difficulty resulting from overcapacity.

By recognizing these different forms of management, fish exploitation must better identify the feasibility of exploitation regimes and break with an illusion of control of the fleets performance and the predictability of the resource, in order to explore forms of regulation of greater complexity [GUN 98, GUN 02, MCG 01, WOR 09].

2.2. Sustainability evaluation methods: translation of scientific knowledge for a political debate

2.2.1. *Classification and good use of the sustainability and responsibility evaluation methods*

IE extended to ecological economics, discussed in section 2.1.2, has generated a proliferation of quantitative evaluation methods of the impacts of human activities covering the totality of the fields underlying the notion of sustainability, and then of social responsibility.

Few satisfactory classifications have been produced until now because reflection is not yet stabilized [KLO 12, NES 07].

Nevertheless, the axes of classification can be distinguished as follows:

– use of quantitative data, of qualitative data or of dimensionless indicators (indices), to link with algebraic models on the one hand and multi-criteria analysis on the other;

– material and energy flows balance, including the identification of losses and yields;

– impact and chains of cause and effect evaluations by sets of agglomerated indicators, capable of reducing scientific knowledge to a few synthetic indicators;

– integrated mono- or multi-criteria approaches combining environmental, ecological, economic and social criteria;

– product-, process- or service oriented methods relating the data of the attached system to a functional unit;

– lifecycle approaches or approaches based on the identification of a chain of "from cradle to grave" transformations, linear or in cycles;

– partial approaches that only consider a part of the extraction process, production, consumption and production of waste and emissions;

– *ex-ante or ex-post* balance methods, some assessing the state of a situation in relation to past data, others allowing projections and a certain prospective.

Historically, the elementary elements have been fixed since the 1960–1970s by methods of material and energy flows balance respecting the principles of thermodynamics. Naturally, a concept of lifecycle emerged when people were looking to establish the flows that would allow industrial and economic production to operate, from the extraction of raw materials, to objects or artificial structures, and then to rejected waste. The environmental LCAs are subject to standardization in constant progress (ISO 14040 and SETAC). The monocriteria balances on fossil carbon (carbon balance), domesticated energies (energy balance) or some toxic products (labels and environmental standards) are well known to the general public. However, trends today lean towards more inclusive methods.

Two major sets of methods are opposed: the quantitative IAMs, those that inspire the works on climate change, and the LCSAs, that propose global or integrated multi-criteria assessments (detailed in section 2.2.2). The concept of integration means that the criteria on the impacts are harmonized to hold a joint and balanced reasoning on the environmental–ecological, economic (in the broad sense) and social (political, human development and prosperity) dimensions. From these analyses, reasonings are derived that combine the criteria of sustainability to the concept of societal responsibility presented in section 2.1.2.4, which refers to the notion of social utility of the activities and political decisions, and of governance.

Most of the methods which consist of building assessment systems rather relate to a form of engineering whose task is to translate the immense knowledge of science and experts. Their aim is to make them coherent through the process of planning and management, or of decision-making under uncertainty and complexity, or at least through political debates and assessments of scenarios and prospective models.

A review of the most recent applications of the LCSA-based methods in fisheries is proposed in section 2.2.2.

2.2.2. *Sustainability analysis of fisheries and aquaculture supply chains: state of the art, trends of the new practices and case study*

2.2.2.1. *State of the art and trends*

We have seen that generally in the industry new accounting tools appear to establish the energetic, environmental and economic balance of products, supply chains, industrial processes, and services, at any system scale (e.g. LCA, AFME (*Analyse de Flux de Matière et Énergie, or analysis of matter and energy fluxes*), carbon balance). These tools are beginning to be used to objectively analyze and compare the opportunities of marine fishery and aquaculture exploitation as detailed below. These approaches are at the origin of the most genuine way to change the global economy, whose current model is in crisis (e.g. new green taxation, development of eco-systemic services and of ecological engineering, productivity of resources for growth, etc.). Above all, they inspire competition among distribution professionals to control costs and to justify prices to consumers. Some companies specialized in the distribution of seafood products are committing themselves in this new direction (as for example, the brand Findus or the Super-U distributors) according to so-called "niche" strategies, based on the traceability of products, their labeling and the use of eco-labels (sometimes used for personal gain, as seen in *greenwashing*). Similarly, a number of NGOs and governments have proposed labels of "sustainable or responsible fishing", generally based on summary analysis derived from the opinion of experts or from reports published by international institutions (FAO, CIEM). These labels are targeting the responsibility of the consumer in their purchasing behavior and to promote their awareness to the issues of sustainability (biodiversity conservation, overfishing, selectivity and impacts of fishing equipment, accessory catches, discards at sea, etc.).

However, these strategies are not yet in a position to influence the global markets. The growing awareness of environmental responsibility (biodiversity and overfishing) at the scale of the consumer, combined with the effects of resource scarcity, are promoting a change in positions of the intermediaries in the supply chain. The mechanisms that govern these changes, and the associated manipulation risks of the consumer, currently lack formal assessment [ALD 08, DAL 98, FAO 08, GUI 08, NEW 04, TVE 06].

In the fields of seafood products and aquaculture, the use of LCA largely dominates the other environmental approaches. This approach is rapidly gaining popularity, as a number of recently published review articles show [AVA 13, HEN 11, VAZ 12]. The application of LCAs in this domain has foremost been focused on the following categories of potential impacts, most of which are linked to the use of fossil fuels: global warming, acidification, eutrophication, formation of photo-oxidants, depletion of the ozone layer, human toxicity, aquatic ecotoxicities (marine and freshwater) and depletion of abiotic resources (fossil fuels, water and minerals). A category of specific impact is often associated: the "Biotic Resource Use" (BRU). Meta-analysis clearly shows that the sources of impacts on the environment are largely dominated by the use of fossil fuels, either directly (fuels for thermal engines or furnaces) or indirectly (thermal electricity), and is true for essentially all the categories of impacts. This applies both to fishing activities [AVA 13, VAZ 12] as well as to those of transformation [AVA 14c, AVA 14e] and, to a lesser extent, to those of aquaculture [HEN 12]. The other sources of significant impact result from the use of anti-fouling paints on boat hulls, which majority influence the marine ecotoxicity category, and from the use of building materials, particularly of metals. Even if the latter are used either in a slightly elaborated form but massively for the construction of infrastructures (boats and factories), or in reduced quantity but in an elaborated form (e.g. chrome steels of engines and copper of electrical systems) and therefore more impacting, they contribute heavily to the following impacts categories: eutrophication of fresh waters, human toxicity, ecotoxicity and depletion of abiotic resources [FRE 14].

The use of these categories of impacts entails certain limitations due to the early stages of the LCA approach, in particular in the areas of seafood products and aquaculture. These weaknesses are in the process of being resolved by two approaches: (1) the use of specific indicators, in this case for fishing or aquaculture, and (2) the consensual adoption of new categories of impacts. With regards to the first approach, which is often only a preliminary stage of the second, it should be noted that the frequent use of indicators is related to the following specific impacts: use of biotic resources, disturbance of the sea-bed, proportion of juveniles, discards at sea and accessory catches [AVA 13]. Regarding the second approach, the proposals of Langlois *et al.* [LAN 14a], unified under the "sea use" category of impact, will be illustrated with the following operational proposals: formalization of the depletion impact on biotic natural resources (I_{BNR}) resulting from fishing at the ecosystem level (appropriation of the primary production) or at the level of the species (recovery time of the removed biomass of the stock) [LAN 14b], formalization of the impact of supporting life functions resulting from the removal of biomass at the species level, from the destruction of the sea-bed (including the often positive effect of artificial reefs) and from shading by floating structures [LAN 14a, LAN 14c].

The assessment of the sustainability of fishing and aquaculture supply chains can therefore be quantified at the level of the three pillars of durability:

– environmental, through a series of LCAs ranging from the exploited ecosystem to the plate of the consumer. These LCAs are either attributional (that is to say descriptive) or consequential, thereby essentially comparative with the aim to predict the consequences of technological choices or of markets;

– social, based either on the social LCA (SLCA), although this approach remains to consolidate [BEN 11], or on extensive sets of existing social indicators such as those proposed by the Global Reporting Initiative (GRI, www.globalreporting.org/). These various indicators may relate to the different stakeholders of the supply chains: workers, consumers, local communities, civil society and other actors. In very general terms, they fall under the scope of employment and human well-being (including health and safety, and, in this case, the nutritional quality of products), of trade union rights, social benefits, employment security, of the absence of child labor, forced labor, various forms of discrimination and corruption, the decency of wages and hours of work, etc.;

– economic, based either on the "Lifecycle Costing" *(LCC)* [HUP 04], although its benefit is expressed mainly in consequential approaches (e.g. compared economical performance of two types of fishing units throughout their lifecycle), or on extensive sets of economic indicators such as those proposed, again, by the GRI. Among the latter, those which can be better associated to a functional unit in the supply chain should be retained. They can involve the following main categories: generated and distributed rent; governmental financial assistance; investments in infrastructure and local services, direct and indirect associated impacts; local provisioning.

Although some large vertically integrated firms of the fishery sector communicate the results of sustainability through reports of the GRI type or quality certification requests – in both cases with a concern for the improvement of their brand image – we do not have knowledge of integrated scientific study of sustainability of fisheries or aquaculture supply chains comparable to those, still very limited, which exist in the agricultural or industrial sector [SEU 08]. The only known attempt of integrated approach covering all three pillars of durable development of a complex set of fisheries and aquaculture supply chains is the one in the process of being completed in Peru[6]. Although it focuses primarily on the environmental pillar, it includes two social indicators and three economic indicators.

2.2.2.2. *General presentation of the case study: the Peruvian anchovy*

Let us now give details about the representative case study of what can be done in the field of study of fisheries and aquaculture supply chains sustainability. It focuses on the evaluation of supply chains derived from the anchovy (*Engraulis ringens,* locally known as *anchoveta*) in Peru, extending from the ecosystem to the shelves of stores.

---

6 It concerns the project *Anchoveta Supply Chain* (ANCHOVETA-SC, http://anchoveta-sc.wikispaces.com), which constitutes a contribution to the joint international Laboratory "Humboldt Current System Dynamics" (LMI-DISCOH) coordinated by the Research Institute for Development (IRD) and the *Instituto del Mar del Perú* (IMARPE), among several other institutions.

Figure 2.2. *Small-scale Peruvian purse seiner (photo Pierre Fréon)*

The Peruvian fleet sourcing anchovy is the largest national fleet targeting a single species at the worldwide level. This fleet consists of purse seiner vessels ranging from 4 m^3 to more than 700 m^3 of hold capacity and has landed on average 6.5 million tons per year during the period 2001–2010 [AVA 14c, FRE 10]. Only about 1% of the catch is intended for "direct human consumption" (DHC), despite the efforts of the government to increase this percentage [FRE 14c]. The majority of the catch is reduced to fish meal and fish oil (FMFO, collectively), and then exported for the feeding of aquaculture or land farmed animals, mainly in Asia and Europe. The fishing fleet is concentrated in two large unit operating under different judicial regimes: the industrial fleet on the one hand and that of the coastal small- and medium-scale (CSMS; Figure 2.2) fleet on the other. The industrial fleet (vessels of more than 32.6 m^3 of hold capacity) is composed of two major parts: (1) around 700 wooden ships nicknamed *Vikingas* from 32.6 to 110 m^3 of hold capacity belonging to individual fishers or to small non-integrated companies (Figure 5.3); (2) 660 steel vessels belonging to large companies vertically integrated, most of which have a holding capacity located between 75 and 635 m^3 (Figure 5.4). The coastal small-scale fleet includes more than 600 boats of less than 10 m^3 of hold capacity, while 780 boats of 10 to 32.6 m^3 of hold capacity constitute the medium-scale fleet [FRE 14a]. The catches of the fleet of steel vessels represents approximately 81% of the total catches of anchovy intended for fish meal, while those of the *Vikingas* represent 19% [FRE 14c]. The landings for the whole of the industrial fleet are intended for the milling industry, whereas the landings of the CSMS fleet are in principle reserved to DHC (fresh, frozen, canned food, salted semi-preserves, salted food), although in practice more than half of the catches end up, in a largely illegal manner, in fish meal factories. Various processes of certification and quality verification are in progress.

Figure 2.3. *Peruvian industrial seiner, wood hull (photo Pierre Fréon)*

Figure 2.4. *Peruvian industrial seiner, steel hull (photo Pierre Fréon)*

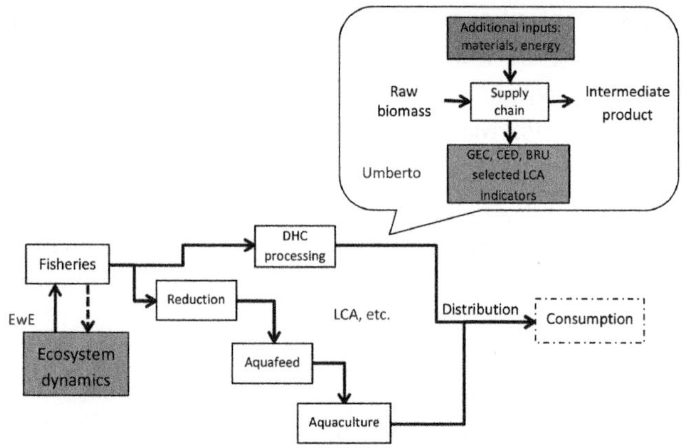

Figure 2.5. *Simplified diagram of the coupling between a marine ecosystem and fishery and aquaculture supply chains (this is only one example showing how the processes and sub-industrial processes are detailed within the production chains; environmental and socio-economical impacts are detailed further on). The names of the main software used for modeling appear in light grey (modified from [AVA 12])*

Various fishing and aquaculture supply chains are competing for the use of the anchovy resource. Anchovies generate a series of environmental and socio-economic impacts, both at the national level (Peru, but also in European and Asian countries where the aquaculture species are cultivated) and worldwide (Figure 2.5). The project aims to understand these dynamics and to quantify the associated impacts in order to assess the level of optimization of the balance between the different Peruvian supply chains derived from the exploitation of anchovies. In order to achieve this objective, a conceptual framework is proposed to evaluate the performance of the sustainability of the systems of sea and aquaculture products, to a high level of aggregation (production sectors). It covers three main phases [AVA 14a, AVA 14d, FRE 10]:

– characterization and modeling of the biophysical and socio-economic flows associated with systems of sea and aquaculture products being studied;

– sustainability comparison of the supply chains using a set of indicators of durable development covering energy, nutrition, ecological, environmental, social and economical aspects;

– determination and simulation of alternative scenarios of exploitation and of anchovy use based on different political options.

Finally, the ultimate goal of the research is to provide strategic and managerial recommendations based on the results of the evaluations/comparisons.

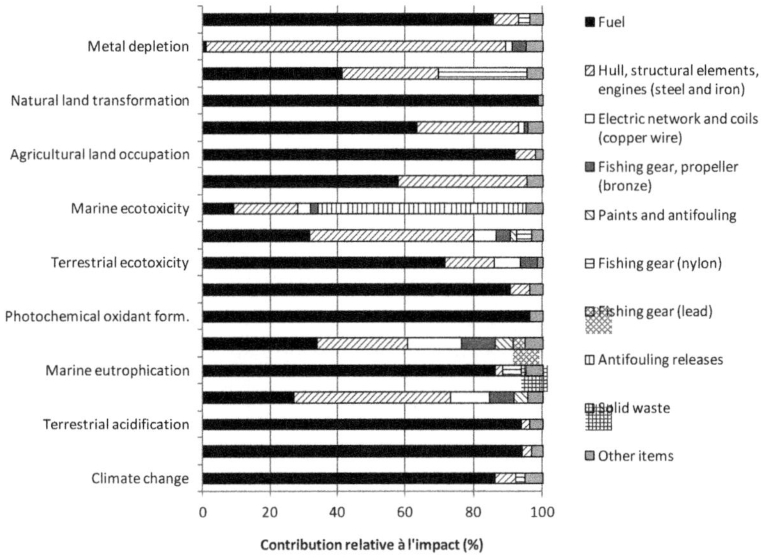

Figure 2.6. *Environmental impacts relative to the main used elements during the lifecycle of the vessels from the industrial fleet of steel purse seiners exploiting the Peruvian anchoveta (according to Table 4 of [FRE 14b])*

2.2.2.3. *Environmental dimensions of the case study*

The accounting biophysics framework of flows and stocks of materials and energy that occur along the supply chain is that of the lifecycle assessment (LCA). The software Simapro©, coupled with the international EcoInvent database [ECO 12], has been used. Here, as elsewhere in the world, fuel consumption remains the main source of environmental impact (Figure 2.6), despite the Peruvian industrial fleet being the one that consumes the less fuel in the world (16 kg) per ton disembarked. This is due to a high abundance of the resource and the high number of landing points along the coast. In contrast, some demersal and tuna fleets in the world consume up to three times more fuel mass than landed fish mass [AVA 13]. Even in the favorable case of the Peruvians seiners, auxiliary and exosomatic energy represents the immense majority (99.87%) of the cumulative energy demand (CED) of the fleets. Among the fleets of anchovy, the industrial fleet shows a performance of fuel use at least twice greater than the average of the CSMS fleets, due to the effect of scale economies. The latter remains however much weaker than expected with regards to the enormous difference in holding capacity of the different fleet segments. The result is essentially an environmental efficiency, on average two times bigger than for the industrial fleet. Within the latter, the *Vikingas* unit is slightly more efficient than the unit of steel vessels because of better fishing strategies related to their independence from the large integrated fishing companies. The latter are in fact constrained by the overcapacity of their fish meal factories [FRE 08, FRE 14c]. Similarly within the ACSA fleets, the unit of vessels with less than 10 m^3 of hold capacity is astonishingly the most efficient of all the fleets, due to the access to the richest fishing areas and to an excellent fishing strategy (Figure 2.7).

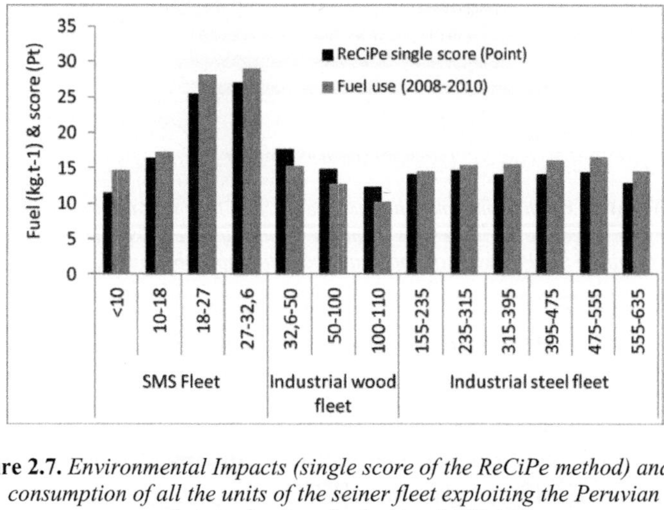

Figure 2.7. *Environmental Impacts (single score of the ReCiPe method) and fuel consumption of all the units of the seiner fleet exploiting the Peruvian anchoveta, by disembarked ton [AVA 14f]*

With regard to the reduction in FMFO, the analysis of the environmental impacts of two Peruvian plants (Figure 2.8) confirms once more the prevalence of impacts resulting from the use of energies, in this case the fuels used for cooking the anchovies and even more so for its drying. The use of natural gas as the main energy source is preferable than that of heavy fuel oil in terms of impacts.

Figure 2.8. *Cyclone separators from a Peruvian factory of fish meal and fish oil capable of handling more than 120 tons of fresh fish per hour (photo Pierre Fréon)*

COMMENT FROM FIGURE 2.8.– Notice the huge size of facilities in relation to the size of a worker (left center of the image) and the cleanliness of the facilities, which is often absent in semi-clandestine factories for the production of residual fish meal.

Within the studied Peruvian systems of freshwater aquaculture – trout, black pacu (*Colossoma macropomum*) and tilapia – aquaculture food (aquafeed) for trouts presents higher environmental costs than aquaculture food for other herbivore or omnivorous species. A similar trend is observed when farming these same species of fish, mainly due to the strong contribution of aquaculture feed in their environmental pressures. Whatever the case, the substitution of small-scale produced food by commercial foods, in spite of the systematic improvement of food conversion rates (FCR), does not always reduce the global environmental impacts. This is mainly due to the additional use of energy which is associated with them. The environmental performance of the ingredients in food is heavily influenced by their processing degree, in particular by the energetic contribution of the specific activities related to the processing of raw material. Among cultivated species, the aquaculture of the black pacu shows the best environmental performance. Taking into account the importance of food on the overall effects of the lifecycle of aquaculture products, the Peruvian industry of aquaculture feed should be preferentially provisioned with inputs that are less refined and, in general, less impacting on the environment. In regards to the fact that the performance of fish farming is not strongly affected, in particular the TCA and the costs structure. For example, Bolivian soybean-based products must be preferred to those of Brazil; fish meal of high quality should be preferred to that of lower quality; protein concentrates should be avoided [AVA 14d].

2.2.2.4. *Socio-economic dimensions of the case study*

The analysis of socio-economic flows offers an overview of the social and economic dynamics occurring in parallel to those of materials and energies. Socio-economic data, such as the number of direct jobs throughout the different supply chains, the production costs and selling price, have been recorded. They allow to calculate the added value and the gross profit (very close to the economic rent because taxes, fishing rights and subsidies are very low in Peru).

The absolute number of jobs provided by the industrial fisheries of anchovy is much greater than that provided by the CSMS fisheries. This is due to the greater development of the industrial sector. However, as in similar case studies, the ACSA fleets generate four times more employment per landed ton than the industrial fleet, more fish to feed the populations and less discards at sea (Table 2.3). The gross profit is more than double for the CSMS fleets of anchovy (74 US$ per ton) than for the industrial steel or wood fleets (33 and 32 US$ per ton respectively). This is essentially due to the lower production costs of the CSMS fleets.

Criteria	Industrial fleet (steel + Vikinga)	CSMS fleet[a] (SS, MS)
Number of fishers employed[b]	19 000 (3 100)	8 800 (2 900)
Number of fishers per landed 1000 t	5	19
Number of fishers per landed 1 Million USD	23	96
Landings for DHC per year (t)	0	132 000 t
Landings for IHC per year (t)	5.2 Million t	324 000 t (SS: 17%; MS: 83%)
Total landed value per year (USD)[c]	1 025.7 Million	91.4 Million
Fuel use per year (t)	84 000 t	11 000 t (SS: 7%, MS: 93%)
Landed tonne per t of fuel used (t)	70 (Vikingas: 124, Steel: 64)	40 (SS: 68; MS: 38)
Weighted LCIA score[d] per landed t ($Pt \cdot t^{-1}$)	14 Pt	23 Pt (SS: 11; MS: 24)
Cumulative Energy Demand per landed t ($MJ \cdot t^{-1}$)	1 890 MJ	6 810 (SS: 3 260, MS: 7 210)
Selected LCIA midpoint indicators per landed t[e]	CC: 67.5 AP: 0.72 EP: 6.7E3	CC: 115 (SS: 52.1; MS: 123) TA: 1.23 (SS: 0.55; MS: 1.31) EP: 10E3 (SS: 5.7E3; MS: 11E3)

a) CSMS: coastal small and medium-scale purse seine fleet landing >10% of anchovy, SS: small-scale (<10 m^3 hold capacity), MS: medium-scale (10–32.6 m^3).
b) In parentheses, jobs expressed in full-time equivalents.
c) Average prices 2003–2013.
d) Single score of the ReCiPe characterization method.
e) Midpoint indicators of the ReCiPe method; CC: climate change (equivalent kg CO_2), TA: terrestrial acidification (kg SO_2 eq), FE: freshwater eutrophication (equivalent kg P). As for FMFO factories, they generate sixteen times less jobs than the canning plants (Figure 2.9) per functional unit, but generate a gross profit substantially higher, for example 19% more than for canned products [FRE 14c]. The use of natural gas in FMFO factories of superior quality (Prime) is not only less impacting on the environment as described above, but also it generates more gross profit.

Table 2.3. *Comparison of industrial and small-scale fleets fishing Peruvian anchovy, including the landings of other species, with mass allocation of economic values. Reference period of 2005–2010, unless otherwise indicated (modified from [FRE 14a], according to Thomson's original design, 1980)*

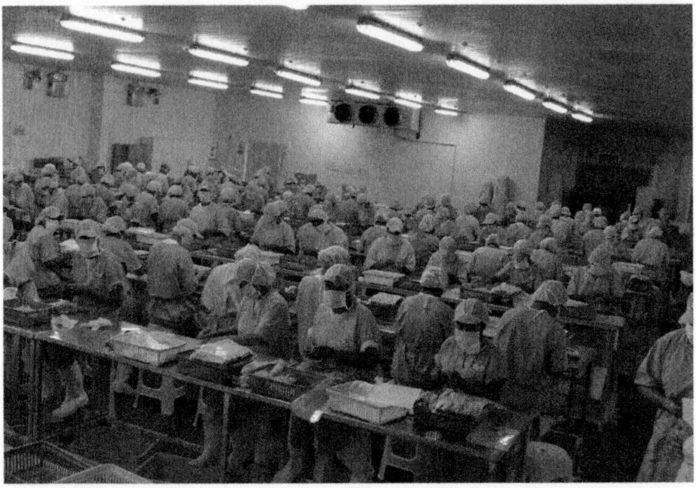

Figure 2.9. *Fillet production line in a production plant of semi-preserved anchovies in Peru, with a current production capacity averaging the 4.7 t of anchovy fillets per day, equivalent to approximately 22.5 t of fresh fish fished 4–6 months earlier due to the process of maturation (photo Angel Avadí and Pierre Fréon)*

COMMENT ON FIGURE 2.9.– Notice the high number of employees (600 direct jobs for the whole plant, divided into two rotating teams) and compliance with health standards (mask, cap), in contrast with the poor health conditions of the majority of the landings of small-scale fishery (not shown here).

The studies of environmental and socio-economic impacts summarized above for the fishing fleets of anchovies and a few examples of plants have been repeated for all the stages of the supply chains:

– processing plants of various products of human food [AVA 14d];

– fish meal and aquaculture feed factories and aquaculture farms [AVA 14e].

2.2.2.5. *Integrated and comparative analysis of the sustainability of full-scale supply chains*

From the link-by-link analyses of the production chains, a large number of environmental impact indicators, of energy efficiency, of the nutritional value and of the socio-economic impacts of sea products have been calculated in an integrated manner throughout their respective chains of production, by expressing them per ton of fish in the product [AVA 14b]. A number of these indicators have been selected in order to present legible comparative tables and graphics for the stakeholders. It concerns the biotic use of resources, the rate of the energy return on investment (EROI), an integrated indicator of the LCA (single score aggregating all indicators), a nutritional index specific to Peru, employment and gross profit (Figure 2.10).

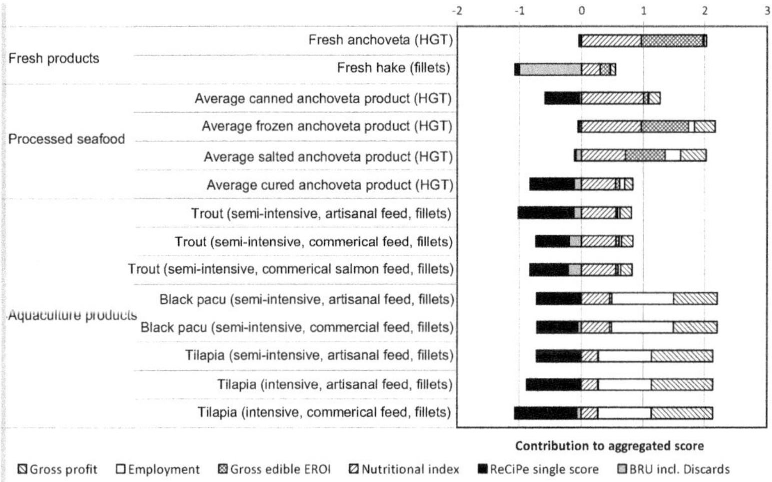

Figure 2.10. *Classification of all the DHC products from the supply chains of anchovy studied rendered at the doors of the factory, according to the set of proposed indicators, per ton of fish in the product (the shortest negative bars and the longest positive bars represent a better performance). HGT = decapitated, gutted, tailed [AVA 14d] (see color section)*

COMMENT OF THE FIGURE 2.10.– BRU (biotic resources use) represents the use of biotic resources, including discards; EROI represents the ratio between industrial energy and nutritional energy.

More detailed indicators have been added to these enlightening indicators such as those provided by the methods of conventional LCA described at the beginning of section 2.2.2, the impacts on the biotic natural resources at the species and the ecosystem level, the costs of production and the value added.

The nutritional index specific to Peru has been adapted from indices well established in the literature. It takes into account the nutritional deficiencies identified in Peru [AVA 14b]. The nutrients retained for the evaluation are therefore the proteins, the Omega-3 fatty acids, the unsaturated fatty acids (EPA + DHA), the other unsaturated lipids (including the Omega-6 fatty acids), vitamin A, vitamins B-12 and D, calcium, potassium, and phosphor, iron (beneficial nutritional substances), and sodium and the saturated fatty acids (nutritional substances to limit).

With respect to the environmental impact of products for human consumption based on anchovy, the finest products (canned products and semi-preserved) represent a greater cost than the less refined products (salted food and frozen). In addition, the highly processed products with a high energetic intensity (canned anchovy or semi-preserved) represent a heavier environmental cost than the less energy-intensive products (salted and frozen anchovy, aquaculture products). This trend is confirmed when all the products are compared in relation to their EROI, that is the ratio between industrial energy and nutrition energy.

The trend is similar for the other dimensions of the analysis: the products from salted and frozen anchovy generate less jobs and profit than preserved and semi-preserved, which generate much less than aquaculture products (Figure 2.11). In addition, derivatives from anchovies have better nutritional properties than aquaculture products. On the whole, the less energy-intensive industries (freezing and salting of anchovies) are less satisfying in terms of economic impact, but offer a better nutritional value and a lower environmental impact than the other industries. The aquaculture products maximize profit and the creation of jobs, but present poorer energy efficiency and a nutritional value lower than the products based on anchovy for human consumption [AVA 14b].

Figure 2.11. *Semi-intensive aquaculture basin of black pacu (Colossoma macropomum) in the Amazon region of Iquitos, Peru. The average size of the basins is 0.86 ha, the production in the order of 10 t/ha/year, the production cycle of 12 months for an individual of average weight of 1.2 kg and the employment of 1 person/t (photo Angel Avadi) (see color section)*

Preserved products are sometimes preferable to improve the nutrition of vulnerable (and often remote) communities in Peru, due to their long life duration and their easy transportation and storage. However, their high price and the preferences of consumers challenge this advantage. The dilemma could be resolved by the implementation of a cold chain for the fresh and frozen fish, despite the increase of impacts on the environment, in particular if the chain extends to the mountainous areas of the country. Another solution would be to improve the current system of conditioning by replacing the metal boxes with plastic or composite packaging to reduce the costs and the impacts on the environment of the production and of the transport [AVA 14c, FRE 14c].

2.2.2.6. *Comparison of exploitation scenarios for the case study*

The scenarios of exploitation have been defined by the coupling of an ecosystem model, Ecopath with Ecosim (*EwE*, [CHR 04]), to a model of material and energy flow (Umberto©, a MEFA tool described in section 2.1.3.1.; www.umberto.de/en/). The coupling of models has been designed as a means to identify and analyze the ecosystem/anthroposphere interactions. The *EwE* model adjusted to the historic chronological series of biomass and catches of the main fishery resources of the system of the North Humboldt currently consists of 32 functional groups [TAM 08]. The simulations of scenarios have been executed over a period of 30 years commencing after the historic period. The studied scenarios (Figure 2.12) include a projection of the *status quo* (scenario 1), an evolution toward more human consumption to the detriment of the FMFO production (scenario 2), and a scenario of radical diversification which reduces by half the fishing mortality of the anchovy in order that other fish stocks are restored and can be exploited for DHC (scenario 3). The implications in terms of biomass associated with these different scenarios are shown in Figure 2.13.

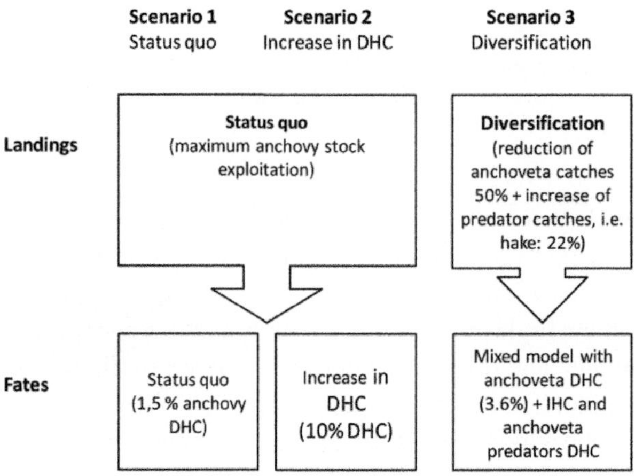

Figure 2.12. *Alternative exploitation scenarios [AVA 14d] (DHC = direct human consumption)*

From the coupling results of the models, the same headline indicators that previously have been calculated for each scenario, per functional unit of sea and aquaculture products (see Figure 2.10 for the reference situation), to which some global indicators of the state of the ecosystem proposed by the IndiSeas project (www.indiseas.org/; [SHI 10]) have been added: mean trophic level of landings, percentage of predators in the landings of commercial species, inverse of the fishing pressure (Figure 2.14). The comparison of the scenarios shows that scenario 2 presents a better contribution in terms of sustainability in Peru, due to a better compromise between the environmental, social and economic impacts (positive or negative), determined by the increased share of products for human consumption (Figure 2.15). However, such an increase would require taking up the challenge of finding commercial opportunities for these products, ideally in large part in Peru, but also, and necessarily, on the export markets. Scenario 3 shows poor performances, but deserves further exploration: simulation of a less drastic decrease of anchovy catches, better taking into account of changes related to species other than the two emblematic species (anchovy and hake) of the ecosystem, in particular the other commercial species (Figure 2.13) and the species having a potential for tourist exploitation.

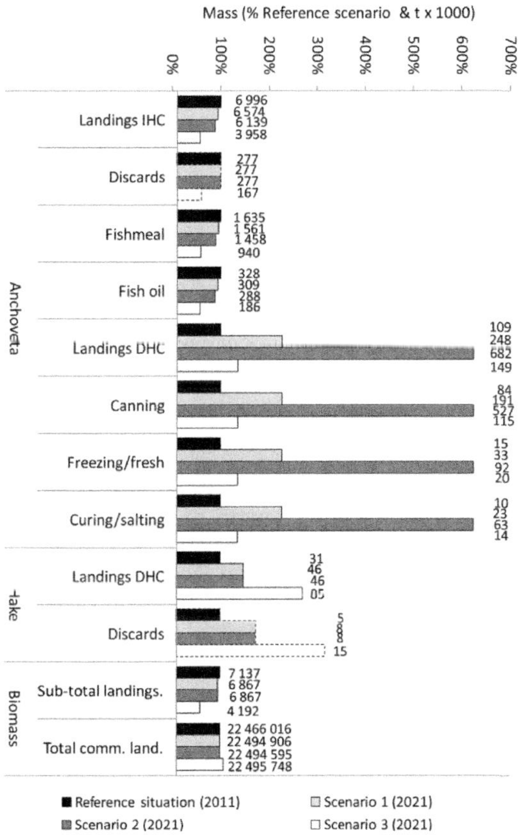

Figure 2.13. *Main biomasses associated with modeled scenarios, in thousands of tons. The percentages reflect the variations in relation to the reference situation [AVA 14d]*

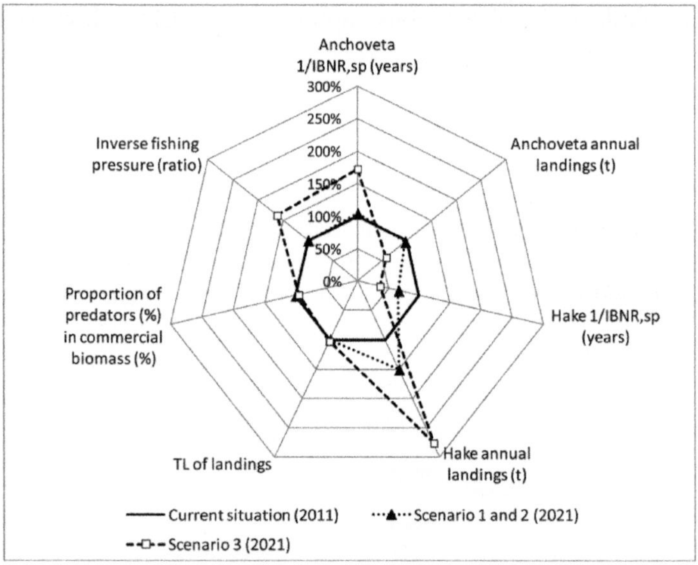

Figure 2.14. *Indicators of the impacts on the Northern ecosystem of the Humboldt current as a function of the different exploitation scenarios. IBNR, sp = depletion impact on biotic natural resources at the species level; TL = trophic level [AVA 14d]*

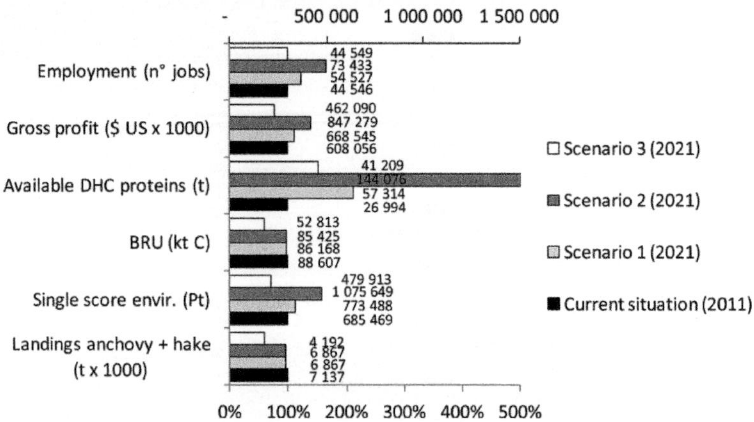

Figure 2.15. *Comparison of the main indicators of the environmental and socio-economic impacts summed up from the whole set of products of the production chains linked to the anchovy (fishing and aquaculture) and hake (fishing) in Peru according to the different scenarios of exploitation (modified according to [AVA 14d])*

2.2.2.7. *Recommendations related to the case study*

A number of management recommendations to improve the performance of supply chains based on anchovies can be envisaged [AVA 14a, AVA 14d, FRE 14a, FRE 14c]:

– in the case of DHC, improving the insulation of fish holds on board vessels and enforcing the use of ice (or conservation techniques of substitution) for vessels landing fish. This measure would allow improving the quality and increase the quantity of landed anchovy for this supply chain and which would really benefit from a transformation into products for direct consumption. In addition, this would allow improving the sanitary conditions;

– raise awareness in fishermen and in control staff about sanitary issues. This awareness would be likely to increase the consumption of fresh fish, including that of anchovies;

– construction/optimization of landing infrastructure and public docks for the CSMS fleet, always with the aim of improving the sanitary conditions of fish handling and stimulating the production and DHC of products;

– limit the practices of illegal transformation of the fresh fish into residual fishmeal by some factories in place of residues and waste from the processing plants, the residual fishmeal having a lesser food efficiency (fresh fish/fish meal ratio) than the other qualities of fishmeal;

– develop cold chains for sea and aquaculture products in order to improve the consumption of fish in Peru, in particular in the highlands and isolated communities;

– promote the use of preserved anchovy containers less impacting on the environment;

– minimize the impact of certain segments of the CSMS fleet through the increase of technification (for example, electronic equipment) and the use of alternative construction materials. It remains that, in total absolute value, the CSMS fleet is the least impacting since it consumes much less fuel per year (but obviously generates much less landings than the other fleets);

– as a result of the aforementioned: prioritize the improvement of the environmental impact of industrial fleets, in particular in the case of the steel fleets, whose dominance in the landings is unlikely to change radically (generalization of the installation of electronic injection engines of which a study of lifecycle costs shows that it is slightly more profitable than the use of conventional engines);

– improve aquaculture practices through the implementation of better management practices, in particular by the small-scale producers, with the objective of optimizing the food conversion ratio (FCR) by means of better feeding management (e.g. calculation of daily rations, differentiated food diet according to the development stages). A global approach combining a better management of aquaculture and of quality food (that optimize the nutritional characteristics and the environmental performance) is desirable;

– encourage the use of agricultural ingredients less harmful to the environment (e.g. maize and local rice, Bolivian soya meal instead of Brazilian) in aquaculture feeds in order to improve their environmental performances. The same goes for the low integration of extremely refined products (e.g. foods incorporating gluten, extracts of proteins, vegetable oils with high impacts related to the transformation of natural land, to the extent where the FCRs are not compromised).

Recommendations about fisheries policies are also proposed, of which some are based on the critical analysis of the literature and represent the consensual opinions of scientists and other analysts in Peru [AVA 14b, AVA 14c, FRE 14a, FRE 14c]:

– assign thresholds of flexible tolerances for the discards of various processes of transformation factories for DHC, according to their specific quality requirements. The threshold currently set at 40% for all landings destined for DHC is arbitrary, because the different industries of DHC have different levels of adaptability to variations in quality of raw materials. The measure would also contribute to reduce the size of the industry for the production of residual fish meal illegally provisioned with fresh fish;

– allow the development of the supply chain of fresh anchovies. Currently, the ships of the CSMS fleet are required by law to deliver fish to the processing plants for DHC only. A legalization of fresh/chilled direct sales, currently limited, is therefore necessary; this is in addition to the improvement of infrastructures and of the distribution;

– homogenize and rationalize the legislation on fisheries, in order to allow a true ecosystemic approach to the management of fisheries, resolving the existing problems associated with the existence of several schemes for the different fishing fleets of anchovy.

2.2.2.8. *Conclusions*

The LCA approach is now sufficiently mature and operational to be easily applied to the field of fishery and aquaculture supply chains. The tools, the databases of environmental information and the accumulated experience allows us to obtain quick results. Even though it concerns potential environmental impacts, estimated very crudely, these limitations are not too problematic when the pursued goal is not to obtain a quantification in absolute value, but rather to:

– identify the main sources of impact during the lifecycles of the various components of a production chain, and more generally the most impacting processes (*hot-spots*) within the chain;

– compare the impacts of different industries;

– compare various scenarios of exploitation;

– attempt, starting from this point, to reduce the impacts either by technological changes, or in promoting certain supply chains, or still in favoring certain exploitation scenarios. The technological and/or political choices, natural or imposed by the decision makers, will depend on these diagnostics.

In the best case, although unfortunately not the most frequent, a convergence is observed between the resulting conclusions of the indicators analysis of the three components of sustainable development (environmental, economic and social). In our case study, this has happened with the option to replace heavy fuel with natural gas as the main energy source in the milling industry. The same has happened when conventional engines for boats have been replaced with electronic injection engines (unpublished results of the Anchoveta-SC project). In these two cases, the change provides both a substantial environmental benefit and an increased gross profit, with the social impacts remaining largely unchanged. However, when it comes to the choice of supply chains or of scenarios to promote, the performances of the different indicators are often opposed between the three dimensions of sustainability. The most frequent case being that the superiority of an environmental option is associated with the inferiority of the economic and/or social indicators, and vice versa, as in the case of the products for human food based on anchovy. Therefore, it is necessary for the decision makers to find compromises in function of the objectives, and above all, in the priorities for the policies in the environmental, food (security and quality), social and economic fields.

2.2.3. *Relevance of the concepts of ecosystem-based fisheries management, bioeconomics, sustainable and responsible consumption of seafood products*

The concept of EBFM and EAF proposed by fishery scientists (see Chapter 3 and Chapter 4 of [MON 14a]) to inspire a different organization of fisheries than the initial approach, more respectful of the exploited ecosystems, shows problems in establishing itself as new management model. Ecosystem-based management is based on an ambiguous definition of ecosystems. If the expression ecosystem-based evokes only the exploited system, then the management remains close to the initial approach, despite the restrictions imposed on fisheries. If it evokes the exploited system with fleets and an arbitrary allocation of what sustains them, then the ecosystem approach is very partial, as we can demonstrate through our case study (section 2.2.2). If the adjective evokes the full interaction between the exploited systems and exploiter, then the models of EBFM/EAF have legitimate concerns to expand, in particular through an adaptive or an explicit evolutive (panarchies) dimension. Through GE, it could be estimated that the interactions of human societies with the exploited natural ecosystems are too recent to have undegone enough feedbacks and to acquire the sustainable/durable properties, or disappear (panarchic cycles). Given the magnitude of changes caused by the industrial age and human demographic explosion, it can be assumed that the return and adaptation time can be long and can compromise the general balance of the biosphere, in other words the conditions for life. Nonetheless, the new approaches of GE or of the new ecology can help us to verify if a partnership is possible between exploited and exploiting systems. The ecosystemic management of the oceans can only be defined if the nature of the relationship is clarified with the exploiting system of human societies as a whole. Regardless of the approach, the objective for human societies is to define how virtuous and durable relationships can appear between the exploited (ocean) and exploiting (human economies) systems by using parameters and accounting rules appropriate to the nature of the systems at stake.

One of the disadvantages of the EBFM/EAF seems to reside in the fact that it does not fundamentally question the economic principles of the opportunities of downstream fisheries in the chain. The bioeconomic approach to fisheries attempts to bridge the gaps of EBFM/EAF by exploring the question of opportunities through the formation of prices and demand (see Chapter 3). Nevertheless, bioeconomics does not sufficiently enlighten managers about the states that the regulation of fisheries could take to agree with the ecosystemic approach when the uncertainty and especially the uncontrollability of the fleets are high. Yet, Chaboud's presentation (see Chapter 3) shows the state of the art and subtleties recently introduced into the models. Research could explore ways to integrate, into bioeconomic models, the parameters provided by the comprehensive evaluations of the supply chains by LCSA. The main point would be to take into account the indirect effects (externalities in approach of ecological economics, ecosystemic services, hidden costs) of human activities on the ocean system (see Chapter 4 of [MON 14a]) and of the macroeconomic and social factors which can manage the policies of fisheries (bottom-up approaches to supply chains). A new economy of resources has been proposed by the green economy to rethink the world supply of proteins and the share that sea products would represent. At the time of the generalization of overfishing and of the full exploitation of the oceans [BRO 03, FAO 12], in spite of certain reconstructions of stocks and ecosystems which are appearing [WOR 09], the search for a new balance between wild resources fishing and aquacultures deserves to be laid out differently. This approach can redistribute the cards of agricultural development, of biodiversity and natural ecosystems protection. It has been shown in several seas that are very rich in biodiversity (Taiwan, Australia and Pacific Islands) that submarine tourism brings more benefits than fishery. In many cases, it could be envisaged to no longer supply low value added activities or with too high environmental costs. The globalization of the economy and the development of tourism, and eventually of eco-tourism, gives opportunities to find new modes for natural resources and ecosystem services valuation. The development of biotechnologies is giving a new vision of natural living resources valuation when the industry understands the limits of oil-based low-cost chemistry. This industrial revolution will also give rise to very innovative models of biodiversity and ecosystems valuation. This will argue to use the ocean as a tank-laboratory rather than for harvesting. There, genetic resources (biodiversity) would be thoughtfully extracted to supply onshore domestic units of intensive production, or biotechnological supply chains with models for the production via chemical synthesis. The concern for bioeconomics is to consider all models which allow to reduce in the long run, the exploitation of biomass food supply well below the calculations of surplus that the initial fishery approach proposes, in order to integrate the constraints of uncertainty and uncontrollability, without losing in economic efficiency with regards to human populations or the concerned countries. The procurement of biotechnological (substances, materials and processes in plants) and agri-food supply chains with high added value (pharmacy, food supplements, functional foods, see Chapters 2 and 3 of [MON 14b]) can gradually compensate for shortfalls in the reduction of massive extraction, scarcely selective and poorly effective (industrial metabolism), of biomass which prevails today. As such, bioeconomics will be able to renew the meaning of handicaps given to the presumably ecological and sustainable deployment of fleets, based on economic criteria, but also that of law and management criteria, acceptable by all actors to a certain regional scale. We will resume this idea in the development of prospective scenarios in section 2.3.

In order to rearrange the fishery supply chain, it is certain that it will be necessary to link the reflections on the effectiveness of the use of sea products to the world food consumption and the impacts of fishing on the resources. The same goes for aquaculture. The combined use of ecosystem-based fisheries management (naturalist logics), bioeconomics (toward a management of provisioning under GE) and the development of responsibility and sustainability logics of consumption, seem to be the only possible approach. It will have to find relevant rules (overall performance and integrated, complex regulations) for the allocation of resources extracted from the oceans. For example, many regulatory measures can be thought of as based on a new and thorough reorganization of the combined choice of fishing gears (more selective), of the power and guidance of the fleets (handicaps), of the periods and duration of fishing seasons, of the fishing quotas (volumes and property) according to the opportunities, and the expected performance to supply the downstream economic supply chains. It must been taken into account that the latter are defined by the markets and the strategies of agri-food production, of biotechnology and of food security in the case of very poor coastal regions without alternatives. This is the implicit approach of the FAO and experts in food consumption or in human development (within the meaning of the United Nations), in particular to find a solution for the provisioning of protein to an expanding human population.

2.3. Interpretations of the evaluations of sustainability/responsibility in global performance: scenarios of complex organization policies and strategies

2.3.1. *Interpretation of the sustainability analysis and knowledge management*

As we have seen, sustainability analyses based on LifeCycle Sustainability Analysis (LCSA) are integrated (environment, ecology, human health and societal usefulness). Nonetheless, they are not scientific models, but rather tools used to support decision-making where research focuses only on the methodological aspect. As in the management of a business or of the economy of a country, the accounting policies can take many variable forms, depending on the political and cultural history. Let us emphasize that the sustainability analyses based on quantitative data of material and energy flows, on impacts and costs (financial, energetic or moral) assessments, are merely simplifications of a more complex reality, according to conventions which must be explicit and rigorous. These systems of knowledge and information management are greared to serve a political and strategic demonstration, then to be discussed. Thus arises the problem of the interpretation of the results which depends on the accounting policies adopted, on the choice of indicators and their agglomeration, and finally, on the meaning that these tools would have for the stakeholders of the political debate and of the resulting projects.

The criteria of sustainability allows to judge products, processes, services, systems, supply chains and even the whole socio-economic system if needed. By choosing a solution, the political actors must compromise between many criteria which may be contradictory. Analyses of the integrated type, particularly LCSAs, are sufficiently rich in criteria to raise the challenge of a generation of new complex type scenarios, sometimes counter-intuitive and generally transversal between compartmentalized academic disciplines.

LCSAs, which provide solutions in the form of expert recipes or synthetic notes, are admissible only if they do not cut short the discussion, but rather fuel it. The analysis experts must be capable of justifying all their resulting simplifications, weightings and presentations. To be effective, analysis of multi-criteria sustainability must compare objects or systems, technological alternatives or assumptions, and help to generate scenarios and solutions without dictating any logistics in the decision-making, as illustrated in section 2.2.2. Choosing solutions is a matter of reformulating objectives strategies, social acceptability, and efficiency (costs), and therefore of policies and trajectories.

The applied cases of the fisheries of Peruvian anchovy and the globalized economic chains they supply (section 2.2.2), illustrate in detail how a large number of fishing policies have not been able to satisfy the three categories of environmental, social and economic criteria. These cases also illustrate that it is by working with more finesse (quantitative multi-criteria approaches based on impacts), based on the comparative analysis of the supply chains, that we can begin to identify the policies capable of taking up the challenge of finding a virtuous, efficient, and globally sustainable organization. In the case of Peru, as for many less industrialized or poor countries, the arbitration between the different ways of valuing the resources will have to find an optimum compromise between several options:

– serving industrial export supply chains poorly effective in global performance;

– serving the cause of local or national food security (a quasi-nationalization or economic domestication of the common good);

– changing the exploitation model (extensive and with high added value), in order to radically preserve the natural ecosystem while ensuring a significantly rent in global performance.

The experts, who lead the sustainability analyses, the generation of scenarios and their comparison, gains credibility by not participating in political analysis organized downstream from their contribution. They must restrain themselves to simply provide input and insights on their methods and results. It is for these reasons that the experts, and even consultants who perform the translation and the simplification of the available knowledge, are not legitimate individuals within the circle of stakeholders of the political debate. When these experts are also (or have been) scientists, which is frequent in fisheries management, it is in their interest to adopt a stance which clearly separates their accounting role from their commitment as scientists to the defense of ecosystems or exploited species. In the case of an overlap in their roles, it is easy for stakeholders who want to manipulate the statistics and the scenarios to discredit these sustainability accounting experts. These questions can renew the precise definition of the interface of roles between the scientific experts and the political leaders who are responsible for making the decisions.

2.3.2. *Large prospective scenarios on the organization of fisheries and marine aquaculture supply chains in a globalized context*

Although we will not develop them here, it is possible to define theoretical scenarios about the evolution of fish exploitation that have learned from overfishing and from the threats related to modifications in the biological and social environment.

Prospective methods are not always very explicit about the construction of scenarios. Building scenarios can have the aim of quantitatively predicting the evolution of variables calculated over the past, which is what the IAMs do, as mentioned in section 2.2.1. When the subject is complex, the scenarios aim to stimulate a political debate and to overcome the usual balance of power by assessing information and knowledge, or even by integrating uncertainty. The construction of scenarios can envisage all the reactions of the stakeholders over short or long periods. The exercise is largely intuitive, because no model enables the mechanical construction of scenarios that create triggers in people's minds. It is based on the identification of different reasoning logics on the same subject.

This scenario, commonly used in negotiation or in the strategic thinking of enterprises or states, has an informal side. The method consists of arranging all the possible and plausible scenarios, through extreme scenarios. They will be able to make the stakeholders react by means of a rejection effect, in order to refocus the debate toward nuanced scenarios and eventually complementary in time (method of contrasting scenarios). These extreme scenarios are based on a "baseline scenario", that is to say one of the most likely to take place, at all the moments where choice imposes itself, following tendencies registered in a situation of origin, by forcing the features of divergences on the diagnosis of the present situation. Overfishing produces both a belief that things can regenerate themselves, as a form of denial, and a sense of alarm which should generate a change of behavior. The situation of overfishing, indisputably a global phenomena presents two cases. One, pessimistic, results in an ocean which will be emptied of exploitable species (direct effects of fishing), will see its biological productivity collapse (indirect effects) and will be populated by a few invasive species without predators, such as jellyfish (see [MON 14a]). This catastrophic scenario insists on the irreversible nature of the direct impacts of fisheries but also on the indirect impacts of other global anthropogenic changes and the implications that are predicted. Conversely, it is optimistically possible to picture that the ocean ecosystem embodies significant resilience (capacity to regenerate) and will be able to provide different resources, even if fishing pressure has to be temporarily reduced. However, both these two extreme scenarios assume a drastic reduction of the fishing pressure, over several years, with the logical economic and political consequences. They will therefore be rejected by stakeholders of the exploitation industries. What the scenario-based approach must provide is a debate on the modification of policies, primarily local or regional, through rupture or through progressive adaptation, depending on the case.

With regard to fisheries, three dimensions of the results of sustainability analysis can orientate the process during the development of scenarios of a sustainable and responsible, sober, clean and prosperous economy: (1) the integrated approach (multi-criteria and in global performance); (2) the GE approach enabling improvements in top-down and bottom-up logics along the supply chains and an adaptive approach (panarchy) and (3) the differentiation of wordings, objectives and regulations according to the metabolic characteristics of fleets (energetic/technological dependence and intensity of human labor). The integrated dimension ensures the definition of criteria of global performance with regards to durable development. GE makes it necessary to explore the objectives of sustainability relatively to a global accounting perimeter of fishery supply to human societies and to learning panarchic cycles. Finally, the differentiation of wordings, objectives and regulation solutions must take account of the balance of global performance of the exploiters activities. One of the principles that our case study demonstrates is, for example, the higher energetic cost of small-scale and, even more, medium-scale fleets, whereas the industrial fleets capable of investing can make significant progress in order to reduce the environmental impacts. Generally, public policies will have to arbitrate measures in order to, on the one hand, minimize ecological and environmental impacts, but especially to adapt fishing restrictions to the economic and social realities, therefore political, of communities of fishers or of small human societies that they supply in short circuit. Then, authorities may prefer to assign as a priority resources to poorly performing supply chains with regards to the environment (energy, fleets maintenance), moderately impacting ecology, but economically efficient (multiplier effects, food security, etc.). At the extreme, some authorities or their communities (human societies) could renounce the collective exportation of an industrial explotation and its rent, in favor of promoting a different productivity of resources into the local (or national) economy. A wide range of fisheries governance may appear if fishery and marine aquaculture issues are globalized in the economy of supplies.

Of course, the scenarios produced by the interpretation of sustainability evaluations of supply chains may follow one another, cohabit or complement themselves according to the trajectories of the states or the regions that implement them, along the way, or on a voluntary basis. To summarise, we can propose the following scenarios to understand the possible trajectories of fisheries and aquaculture management.

2.3.2.1. *Baseline scenario*

It must give account of the overfishing and the overcapacity mechanisms of the fleets; global change represents a number of risks, but also a certain economic and political model in the organization of provisioning. In this scenario, the fishing industry fails to find its regulatory mechanisms, nor is it capable of dissuading illegal fisheries (controllability). Overfishing is taking radical and irreversible forms (species extinctions), and perturbations of the factors of biological productivity (climate change and pollution) accentuate this situation. Aquaculture becomes profitable and develops itself in spite of the negative impacts on the coastal environment, but without contributing to the rescue of the threatened exploitable species. The surviving industrial fisheries feed a niche supply chain and are confined to a few regulated resources, ensuring sufficient volumes. For the most part, the supply chains are reduced to small-scale, semi-professional, subsidiary and are not labour-intensive, which is also casual or opportunistic. Panarchic cycles are increasingly shorter and unpredictable, without allowing a significant replenishment of resources (extreme scenario of the irreversible decline), nor compatible with a long reconstruction cycle (partial, spread over several decades without significant fishing pressure). This reconstruction implies that overcapacity does not perpetuate the situation of overfishing. According to this scenario, human societies would not learn from the current situation.

2.3.2.2. *Scenario of structured, ecosystemic and bioeconomic management guided by experts*

Exploitation follows a top-down logic along the supply chain without converging with the weaker bottom-up logics (logic of responsible consumers and distributors consumption, or of incitement through pigouvian[7] taxes or food security policies). The market for seafood products therefore allows demand and prices (rising) to impose their rules, which results in an intensive aquaculture which is as of yet unable to compensate for the decline of marine resources. Regulations are struggling to be respected, especially in the case of trafficking (stimulated by prices and scarcity), because they are costly to states. The actors at the end of the chain (in contact with the customer) manage to capture added value poorly reflected upstream (fishers), unless fleets integrate a growing part of the

7 Tax paid by the polluter and equal to the cost of the caused environmental damage.

processing chain and eliminate intermediaries. Fishing focuses on lower volumes, intermittently extractable. It remains industrial, international and opportunistic in order to seize fishing right concessions from countries whose resources are temporarily replenished. Locally, ecosystem-based management allows to save some resources with the help of radical restrictions (controllable) and protected areas (PMA). These local measures lead to effective solutions for controlling niche supply chains, no longer industrial but now of small-scales. They are of high added-value or intended for strategic consumption (local rights, food security), yet are in competition with trafficking, which plunders the resources. The world's supply of animal protein is not being challenged and demand is guided by competition effects between concurrent supply chains on the world markets.

2.3.2.3. *Radical scenarios, seizing the political opportunity of the effects of overfishing and of the economic crisis*

These scenarios are focused on the political levers of change. The aim is to acknowledge the situation of overfishing and the impacts that will affect the global procurement of seafood products. A series of factors can give rise to major spontaneous changes in the chains of economic value which guide fishery exploitation, of which half of the fishing capacity may be qualified as traditional-small-scale and the other half as industrial [FAO 12].

These factors are of several orders:

– the market logic of fish and aquaculture supply chains (e.g. fleet overcapacity and increase in operating costs);

– the multipolar economic rivalries which allow poor countries to better negotiate their rent on exported resources;

– the issues of food security (difficult to finance the fight against hunger, the development of policies supporting development through private initiative, the local markets and microfinance);

– the development of biotechnology and of valuation chains of the substances derived from living systems (products from living matter and with very high added value, biomimetics, non-destructive extraction, nature–factory–laboratory approaches).

Although unusual in appearance, this list supplies innovative scenarios of responding to the decline of irresponsible, reckless, ignorant fishery practices and which comply with the tragedy of the commons [HAR 68].

Let us distinguish three forms of radical scenarios:

– scenario of structured exploitation: consists of capping the capacity of fleets well below the calculations of MSY by following a reactive, adaptive (uncertainty about the resource) and community-based (self-control by an empowered community) management. The term "structured" means that fish extraction follows the rules of selective catches, characterized by the complexity of the ecosystem and its productivity. Applied to aquaculture, the term covers the extensive aquaculture on large natural areas, including integrated aquaculture practices such as rice-based aquaculture, where herbivore or omnivore species are cultivated in low density, with little artificial feeding. These aquacultures, particularly small-scale ones, allow marketing less refined sub-products, both of fish and of farming origin, and even from closure of the cycle, but with associated health risks). They can still be harmful to the marine environment, such as when the rearing in basins is carried out from an excessive collection of fry;

– scenario of laboratory-tank oceans for bio-industries: recognizes the new supply chains of biodiversity valuation in high added value for the renewal of domestic strains of intensive industrial agri-food, pharmacy, and biotechnologies. Promotes the prospect of the development of production of animal proteins (seafood products but also insects, etc.) in industrial factories that are more sustainable than the current supply chains of red meat or of farmed salmon, and brings about bottom-up solutions to the problems of the regulation of fisheries by radically reducing the worldwide demand in proteins of wild origin. This scenario allows to fix the valuation on prices and volumes which do not require massive seafood extraction. The ocean life can even be copied or drained without reaching the structure of the biomass, because the production of biomass would be onshore and in artificial and industrial and sustainable eco-friendly systems, supported by clean and renewable energies. This concerns a domestication through technologies (biotechnology and ecological engineering) and a development of intensive onshore production rather reserved for species with high productivity (proteins and other) which can supply the industrial agro-food supply chains. In this scenario, the supply chains that reduce fishing in favor of tourist valuation (underwater observation) can also be considered if fish bring more financial benefits when being observed rather than fished;

– scenario by domestication policies of markets (i.e. nationalization of resources destined for local markets): recognizes that allocation of resources as common goods, *sensu* Elinor Ostrom [BAD 98], can resolve questions of food security or stimulate a more efficient domestic economy than that which is sustained by exports at the price of the speculative international markets. Some countries may find it advantageous to nationalize, or to only allocate the fishing rights of part of the resources which were awarded to export distribution channels to more modest supply chains. Such a valuation has a magnifying power on the local economy (in local currency, through jobs, investment and value-added chains), on the budget of states (improved external balance, tax revenues, less shock effects) and on food independence (direct or indirect distribution supply chains of meal and poultry). In this scenario fall the cases where fisheries renounce to collect biomass from the beginning of trophic chains to focus on species at the end of the high valued chain, particularly with regards to touristic places. Energy independence can also be considered if the ocean (or its derivative, aquaculture) provides renewable biomass, micro or macro-algal in particular, well valued in energy production supply chains.

These exploratory scenarios are based on existing experiences which use sustainability analysis (LCSA) to assess their energetic and micro or macro-economical efficiencies. Integrated, and even panarchic, conceptions allow to adapt the schemes of regulations and socio-economic organization to the system constraints, uncertain and more or less controllable (works of the IPCC on climate and global change, national debate on energy transition in France in 2013).

2.3.3. *Conclusion and perspective: toward an integrated sustainable and responsible approach to fishery and aquaculture management*

The global failure in the in the regulation of fisheries as a mode of exploitation of natural resources (here of wild species) is only avoided in some developed countries that have the means to properly monitors their stocks and implement control and surveillance of their fleets, or, in some cases, of community-based management. This is due to the difficult of finding an agreement between the improvements of our understanding, in biology and ecology, of oceans on the one hand, and what the economy dictates in term of procurements of sea products, on the other hand. Incentives exist for the ability to finance the fishing fleets (number of vessels and embarked means) and the creation of aquaculture installations in the marine environment. Faced with the reality of overfishing, regularly cautioned by scientists, the exploiters of seafood products struggle to respect the imposed rules, where the resources could regenerate, when at the same time, the world market is able to pay high prices for its supplies. Trafficking develops as a consequence. The intervention of states in an attempt to give extensions to supply chains, which are legal but in overcapacity, exacerbate the collapse of stocks, and even the destructuration of whole ecosystems.

In conclusion, this diagnostic of the generalization of exploitation and overfishing, as well as of the global changes of the ocean ecosystem, is placed under the scrutiny of the means of sustainability analysis. With regards to the procurement of sea products (through the MEFA and LCSA accounting methods from the ocean to the plate), this reveals three essential points:

– the regulation of fisheries applied solely on the sectors of fish and aquaculture production are no longer sufficient to promote a significant change in fish exploitation, by fully replacing them with structured aquacultures. Future regulations will from now on apprehend the parameters of the complete chain of suppling or provisioning final customer (from the ocean to the plate). The question to be resolved is the antinomy between: i) the biological or environmental constraints (direct and indirect) on the renewal of resources; ii) the environmental impacts generated by fishing and aquaculture activities at the local, regional and global scales; iii) the market mechanisms that direct the demand or the prospects created by the harvesting of seafood products, arming fleets, imposing the transformation of products to the detriment of fresh fish and making regulatory states that establish compromises on poorly insured short-term positions. This principle justifies that traditional marine fishery and aquaculture rationale should be substituted by the rationale of supply to human societies and its model in GE (exploited-exploiting systems, massive transfer of secondary auxiliary energy from one to the other, without adaptive return). Then, fish exploitation will be able to be distinctly inserted in a problematic of global sustainability of human activities, regarding its purpose rather than its means. Although demanding and by no means trivial, the question of the sustainability of fisheries must draw upon the sustainability parameters of the metabolism of human societies in terms of growth, development and adaptation to a modified biosphere, in proportion to the contribution of resources from the oceans. Let us add to this discussion that the sustainability of fisheries must be considered within an integrated vision of the regulation of the other uses of the sea and of the coastline, in such a way that the management of fisheries does not remain their adjustment variable;

– to inspire regulation of supply chains, it is first necessary to assess the global performance of the procurement by human societies of seafood products. We have shown the possibilities and the importance of an integrated or multi-criteria approach (on the three pillars of durable development). It broadens the conventional economic approaches by internalizing the costs and the full impacts (hidden invoices of the activities) and relates them to functional units which have meaning in economics and politics. However, the ecosystem and bioeconomic approaches defined soley on fisheries and aquacultures, suffer from insufficiently challenging the conventional rationale of the current market (liberal, neoclassical, financialized), and from not supporting the legitimacy of state policies to arbitrate the uses of common goods; i.e., the renewable living marine resources;

– the initial models of fish exploitation are at the same time unsuitable to translate the complexity of processes of the resource collapse, and not designed in a manner to integrate forms of learning to human societies. The latter first occur at local levels, and then worldwide, because human economics is increasingly globalized. The models of fish and aquaculture exploitation will have to integrate the knowledge of different reactions of human societies throughout the world into the consequences of overfishing and of global changes of the environment (see Chapter 4 [MON 14a]). The objective will be to build adaptive meta-models which can capitalize on forms of reorganization of fishing supply chains in shortage of exploitable resources, according to the regions of the world, but also on what we have described about the types of fisheries as a function of the uncertainty of the resources and the uncontrollability of means of fishing. Practically, these models would define the cases in which fleets must organize their campaigns according to a phenomenology synchronized with the variations in distribution of the resource, and the difficulties of the control of the fisheries. These new regulations can establish new forms of fishing and of aquaculture less intense in means (auxiliary human energy), brought by more effective value chains (economic) in terms of market and human development policies (food security in some regions). The exploration of new forms of fishing regimes must respond to the theoretical issue which consists of combining the phenomenology (covariance) of fisheries to the complex, fractal and chaotic one of the distribution of the biomass and its biodiversity in natural ecosystems.

A new model of fish exploitation could therefore follow integrated accounting through LCSAs extended to the data of the metabolism of provisioned human societies, and should also follow an adaptive or panarchic principle. The latter recognizes fluctuations cycles at multiple scales of the resource and of the fishing pressure. It also recognizes the weight of the global interactions between the natural ocean ecosystem and that of the human societies whose worldwide network is considerably connected. Similarly to the way which certain large industrial fleets have been moving for a long time toward resources generally far from home ports (which is relative to the power of vessels), and by fishing rights concession between countries or by free access to international waters, this panarchic global model already exists. But it remains poorly guided and regulated because the feedback from overfishing and overcapacity is too recent.

To conclude, structured aquaculture can be considered an option for the partial domestication of resources of the ocean ecosystem, that economic mechanisms will have to valuate by taking into account the global performance of food and non-food procurement. It opens the way to alternative organizations for exploitations currently dominant in volumes and insufficiently selective. The latter would give up their place to extensive exploitations if they keep the ocean or terrestrial aquatic spaces (lakes, rivers, rice paddies) as free support, or intensive ones if they are to be installed onshore or in confined spaces. The supply chains of biotechnologies and of modern domestication of living creatures which are being developed with industrial means could relay conventional agricultural and fish organizations. This would allow to obtain better efficiency, global performance or sustainability of agri-food procurement, but also of pharmaceutical, chemical and material substances.

What we have defined and attempted to demonstrate on the idea of sustainable exploitation of fisheries seems to us be able to be generalized to the exploitation of all ecosystems. The essential frameworks for this research are GE, ecological economics as well as the accounting tools of sustainability and responsibility as a means of integrating knowledge in political change.

2.4. Acknowledgments

The authors warmly thank Christian Mullon, Director of Research at the IRD, for the improvements made to the presentation of section 2.1, and Angel Avadi, doctoral candidate from the IRD and the University of Montpellier-II, for his contribution to section 2.2.2 of this chapter, as well as Sylvain Bonhommeau and Patrick Prouzet, researchers at Ifremer, for their proofreading and constructive comments.

The theses defended in this chapter reformulate, by adapting them, a number of advanced developments of major authors such as Robert Ayres, Ramon Margalef and Serge Frontier to whom we pay tribute. Although at the origin of the theory of GE, R. Margaleff and S. Frontier did not have the time to formalize it in order to encompass IE.

2.5. Appendices

2.5.1. *Appendix 1*

The use of the term sustainability in French is not trivial for the following reasons. Canadians, whose culture is two-sided, have proposed the term *soutenabilité* (sustainability) to replace durable (from *Développement Durable*), which is characterized by the unique criterion of time or duration. It is rather easily shown that non-sustainable systems can persist for a long time, figuratively postponing to a later date the settlement of their debts. The global financially-based economy has been in expansion for thirty years, since emerging countries seized the opportunity to produce what the world needed at low costs. Some progress was made in pollution, with the status of workers and the protection of the environment in advanced economies until the seventies, but these were diluted by bad performances of the economy at the world scale. French-speaking economists have adopted the principles of "strong and weak" *soutenabilité* from anglo-saxon ecological economics. This is to denounce how orthodox, neo-classical and liberal economists intend to integrate durable development (weak sustainability) in conventional capitalist economy, whereas heterodox models breaking with the dominant system plead for prosperity with a strong sustainability [DAL 11, COS 59, FAU 95]. In this sense, the term *durabilité* (durability) in French is quite poor in comparison to the term sustainability which evokes the means, the purpose and the time. In French, the term *soutenable* becomes a professional term when the term *durable* is that of the political world and of the general public. Furthermore, the intentions of durable development have very early raised the question of the responsibility of the human species about the transformation of the biosphere. A considerable debate associates sustainability and responsibility in an attempt to resolve the following issues: the transcript of scientific (rationalities deducted) and popular (traditional experiences, myths and beliefs) knowledge, the ways to project themselves into the future, the intentions (domestications, technologies), the irreversibilities, the risks, the vulnerabilities, the insurability and the usefulness of activities and their impacts. This work of assimilation will be necessary in order for human societies (or communities) to learn to discipline themselves in order to live and evolve on this planet, and within the biosphere ecosystem, in the long term. The issues of governance or of collective action that these terms shouldn't be confused, for which we have exercised ourselves to detail them in a coherent manner in this chapter.

2.5.2. *Appendix 2*

Elements Approaches	Exploited biomass	Fleets and gear	Sailors	Human societies
Initial approach	Primary and secondary productivity, allometries of species/individuals, trophic levels and natural forcings	Energy costs of fishing campaigns and intensity of the material means (propulsion energies and microeconomic ratios); sometimes artificial/ anthropogenic forcings	Intensity of human labor (by microeconomic ratios)	Parameters of economic and social development
EBFM/EAF and Bioeconomy	*Idem supra* and consumable biomass, accessory catches and discards	*Idem supra*	*Idem supra*	*Idem supra*
Generalized ecology	Idem *supra* and secondary auxiliary energies, natural forcing (primary auxiliary energy)	Secondary auxiliary human energy of the segment of the supply chain that manages the extraction and the transformation of the extracted biomass	Human bodily metabolic energy (biomass) and endosomatic secondary auxiliary energies for the pilotage of the fleets	Endosomatic energies of provisioned human societies (corporal metabolisms) supplemented with secondary auxiliary energy (exosomatic) to the *prorata* of the world total

Table 2.4. *Presentation of the energies considered in the various fisheries approaches*

2.5.3. *Appendix 3*

The concept of biosphere is generally synonymous with a planetary ecosystem without industrial humans. The question of the relationship between man and biosphere thus arises both in philosophical, political, but also scientific terms (energy efficient). GE recognizes the system of human societies as an emerging system endowed with a singular metabolism and with a higher hierarchical level, fully integrated in the biosphere. The new biosphere, modified by recent human activities, is in the process of changing (something which may be regretted) as an ecosystem under the influence of a single species. It would be insignificant with low intensities, but the fact that *Homo sapiens* mobilizes as much secondary auxiliary energy as it does, of different forms, means it certainly isn't. The impact of human activities is a factor of evolution of the planetary ecosystem and trajectories, which historically (climates, stoichiometry, distribution of the biomass, changes of biogeochemical cycles) no species had been able to previously modify. Before the industrial era, these global changes were mainly the fact of primary auxiliary energies, essentially abiotic and chaotic. Pollution and the various anthropogenic modifications of the landscape are convertible into secondary auxiliary energies and may, in theory, enter in the calculations of the metabolism of human societies (such as excreta, diffusions or waste). The fact that human societies are capable of changing the course of primary auxiliary energies (natural forcing) at the origin of the life process, must not distract from the fact that the human factor remains consistent with the complexity of the biosphere system, as predicted by theory (new ecology and generalized ecology). Such a definition of the biosphere would fully integrate the emerging techno-industrial human species and its impacts considered, negative on the original ecosystem. It would then judiciously be assumed that the human factor determines an era of the evolution of the planet: "the anthropocene". In recognizing a planetary metabolism (a generalized eco-energetics which renews the concept of *Life Support System* of the Odum brothers; [ODU 71, ODU 07] section 2.1.3.3) able to fully integrate the metabolism of human societies, generalized ecology shifts the question of the man-biosphere relationship and the co-evolutions of two systems toward the question of the continuation of the self-organized complexification (and therefore fundamentally natural) of the planetary system and the ongoing process of life.

2.6. Bibliography

[ADR 97] ADRIAANSE A., BRINGEZU S., HAMMOND A., *et al.*, *Resource Flows: The Material Basis of Industrial Economies*, World Resources Institute, Washington, April 1997.

[ALD 08] ALDER J., CAMPBELL B., KARPOUZI V., *et al.*, "Forage fish: from ecosystems to markets", *Annual Review of Environmental Resources*, vol. 33, pp. 153–166, 2008.

[ALL 99] ALLENBY B.R., *Industrial Ecology: Policy Framework and Implementation*, Englewood Cliffs, NJ, Prentice Hall, 1999.

[ASC 02] ASCHE F., EGGERT H., TVETERÅS R., "Aggregation level and prediction of fishing vessel behavior and productivity", in SHRIVER A., JOHNSTON R. (ed.), *Proceedings from 11th Biennial Conference of the International Institute of Fisheries Economics and Trade*, Wellington, New Zealand, 19–22 August 2002.

[AVA 12] AVADÍ A., FRÉON P., TAM J., "A framework for sustainability comparison of seafood supply chains", *8th International Conference of LifeCycle Assessment in the Agri-Food Sector*, Saint-Malo, France, October 2012.

[AVA 13] AVADÍ A., FRÉON P., "LifeCycle Assessment of fisheries: a review for fisheries scientists and managers", *Fisheries Research*, vol. 143, pp. 21–38, 2013.

[AVA 14a] AVADÍ A., Sustainability of the Peruvian anchoveta supply chains from sea to shelves: towards a new strategy for optimal use of resources, Doctroal Thesis, University of Montpellier-II, 2014.

[AVA 14b] AVADÍ A., FRÉON P., "A set of sustainability performance indicators for seafood: direct human consumption products from *Peruvian anchoveta* fisheries and freshwater aquaculture", *Ecological Indicators*, 2014.

[AVA 14c] AVADÍ A., FRÉON P., QUISPE I., "Environmental assessment of Peruvian anchoveta food products: is less refined better?", *International Journal of LifeCycle Assessment*, vol. 19, pp. 1276–1293, 2014.

[AVA 14d] AVADÍ A., FRÉON P., TAM J., "Coupled ecosystem/supply chain modeling of fish products from sea to shelf: the Peruvian *anchoveta* case", *PlosOne,* vol. 9, no. 7, pp. 1–21, 2014.

[AVA 14e] AVADÍ A., PELLETIER N., AUBIN J., *et al.*, "Comparative environmental performance of artisanal and commercial feed use in Peruvian freshwater aquaculture", *Aquaculture*, vol. 435, pp. 52–66, 2014.

[AVA 14f] AVADÍ A., VÁZQUEZ-ROWE I., FRÉON P., "Eco-efficiency assessment of the Peruvian anchoveta steel and wooden fleets using the LCA+DEA framework", *International Journal of Cleanear Production*, vol. 70, pp. 118–131, 2014.

[AYR 02] AYRES R.U., AYRES L., *A Handbook of Industrial Ecology*, Edward Elgar, Cheltenham, 2002.

[AYR 05] AYRES R.U., MARTINAS K., *On the Reappraisal of Microeconomics: Economic Growth and Change in a Material World*, Edward Elgar, Cheltenham, 2005.

[AYR 10] AYRES R.U., WARR B., *The Economic Growth Engine: How Energy and Work Drive Material Prosperity*, Edward Elgar, Cheltenham, 2010.

[AYR 78] AYRES R.U., *Resources, Environment and Economics: Applications of the Materials/Energy Balance Principle*, Wiley Intersciences, New York, 1978.

[AYR 94a] AYRES R.U., SIMONIS U.E., *Industrial Metabolism: Restructuring for Sustainable Development*, UNU Press, Tokyo, 1994.

[AYR 94b] AYRES R.U., "Industrial Metabolism: Theory and Policy", in ALLENBY B.R., RICHARDS D.J. (eds.), *The Greening of Industrial Ecosystems*, National Academy Press, Washington, 1994.

[AYR 94c] AYRES R.U., *Information, Entropy and Progress: A New Evolutionary Paradigm*, American Institute of Physics, New York, 1994.

[AYR 95] AYRES R.U., "Lifecycle analysis: a critique, resources", *Conservation and Recycling*, vol. 14, pp. 199–223, 1995.

[AYR 98] AYRES R.U., AYRES L.W., *Accounting for Resources*, Edward Elgar, Cheltenham, 1998.

[BAC 10] BACCINI P., BRUNNER P.H., *Metabolism of the Anthroposphere: Analysis, Evaluation, Design*, MIT Press, Cambridge, 2nd ed., 2010.

[BAD 98] BADEN J.A., NOONAN D.S. (eds.), *Managing the Commons*, Indiana University Press, Bloomington, 1998.

[BAR 08] BARTELMUS P., *Quantitative Economics: How Sustainable are our Economies?*, Springer, New York, 2008.

[BEN 11] BENOÎT-NORRIS C., VICKERY-NIEDERMAN G., VALDIVIA S., *et al.*, "Introducing the UNEP/SETAC methodological sheets for subcategories of social LCA", *International Journal of LifeCycle Assessment*, vol. 16, pp. 682–690, 2011.

[BER 04] VAN DEN BERGH J.C.J.M, JANSSEN M.A. (eds.), *Economics of Industrial Ecology: Materials, Structural Change and Spatial Scales*, MIT Press, Cambridge, 2004.

[BER 05] BERTRANS-LANCO S., Analyse comparée des dynamiques spatiales des poissons et des pêcheurs : mouvements et distributions dans les pêcheries d'anchois (*Engraulis ringens*) du Pérou, Doctoral Thesis, Ecole nationale supérieure agronomique de Rennes, 2005.

[BER 99] VAN DEN BERGH J.C.J.M. (ed.), *Handbook of Environmental and Resource Economics*, Edward Elgar, Cheltenham, 1999.

[BEV 57] BEVERTON R.J.H., HOLT S.J., "On the dynamics of exploited fish populations", *Fishery Investigations Series II*, Ministry of Agriculture, Fisheries and Food, London, vol. 19, 1957.

[BOU 87] BOUDE J.P., MORISSET M., REVERET J.P., "Rente et profit en matière d'exploitation des ressources halieutiques", *Cahiers d'économie et sociologie rurales*, vol. 4, 1987.

[BRO 03] BROWN L.R., *Eco-économie: une autre croissance est possible, écologique et durable*, Le Seuil, Paris, 2003.

[CAR 03] CARLSSON-KANYAMA A., PIPPING EKSTRÖM M., SHANAHAN H., "Food and lifecycle energy inputs: consequences of diet and ways to increase efficiency", *Ecological Economics*, vol. 44, pp. 293–307, 2003.

[CHR 04] CHRISTENSEN V., WALTERS C.J., "Ecopath with Ecosim: methods, capabilities and limitations", *Ecological Modeling*, vol. 172, pp. 109–139, 2004.

[CLA 85] CLARK C.W., *Bioeconomic Modeling and Fisheries Management*, Wiley Interscience, New York, 1985.

[COS 02] COSTANZA R., JØRGENSEN S.E. (eds.), *Understanding and Solving Environmental Problems in the 21th Century: Toward a New, Integrated Hard Problem Science*, Elsevier, Amsterdam, 2002.

[COS 03] COSTANZA R., "Early History of Ecological Economics and ISEE", *Internet Encyclopaedia of Ecological Economics* (www.ecoeco.org/), 2003.

[COS 89] COSTANZA R., "What is ecological economics?", *Ecological Economics*, vol.1, pp. 1–7, 1989.

[CUR 08] CURY P., MISEREY J., *Une mer sans poissons*, Calmann Lévy, Paris, 2008.

[DAL 11] DALY H.D., FARLEY J., *Ecological economics: principles and applications*, 2nd ed., Island Press, Washington, 2011.

[DUV 80] DUVIGNEAUD P., *La synthèse écologique: populations, communautés, écosystèmes, biosphère, noosphère*, 2nd ed., Doin, Paris, 1980.

[ECO 12] ECOINVENT, The Ecoinvent Centre portal, www.ecoinvent.ch/, 2012.

[ELS 06] ELSNER W., FRIGATO P., RAMAZZOTTI P. (eds.), *Social Costs and Public Action in Modern Capitalism: Essays Inspired by K.W. Kapp's Theory of Social Costs*, Routledge, London, 2006.

[EPA 06] EPA/SAIC, Life Cycle Assessment: principles and practice, US EPA Pub., May 2006.

[EUR 06] EUROPEAN COMMISSION, DIRECTORATE-GENERAL FOR FISHERIES AND MARITIME AFFAIRS, *Conference on Energy Efficiency in Fisheries*, Brussels, Belgium, 11–12 May 2006.

[EUR 10] EUROPEAN COMMISSION, *ILCD Handbook: Review Schemes for LifeCycle Assessment (LCA)*, Joint Research Centre – Institute for Environment and Sustainability, 2010.

[FAO 03] FAO, “The ecosystem approach to fisheries”, *Technical Guidelines for Responsible Fisheries*, vol. 4, Suppl. 2, Rome, 2003.

[FAO 06] FAO, “Revenue distribution through the seafood value chain”, *FAO Fisheries Circular*, no. 1019, Rome, 2006.

[FAO 12] FAO, *La situation mondiale des pêches et de l’aquaculture 2012*, FAO Publishing, Rome, 2012.

[FAO 99] FAO, “Indicators for Sustainable Development of Marine Capture Fisheries”, *Technical Guidelines for Responsible Fisheries*, vol. 8, Rome, 1999.

[FAU 95] FAUCHEUX S., NOËL J.F., *Economie des ressources naturelles et de l’environnement*, Armand Colin, Paris, 1995.

[FAU 98] FAUCHEUX S., O’CONNOR M., *Valuation for Sustainable Development: Methods and Policy Indicators*, Edward Elgar, Cheltenham, 1998.

[FIS 07] FISCHER-KOWALSKI M., HABERL H. (eds.), *Socioecological Transitions and Global Change: Trajectories of Social Metabolism and Land Use, Advances in Ecological Economics*, Edward Elgar, Cheltenham, 2007.

[FOL 98] FOLKE C., “Ecosystem approaches to the management and allocation of critical resources”, in GROFFMAN P.M., PACE M.L. (eds.), *Success, Limitations and Frontiers in Ecosystems Science*, Cary Conference IES, Millbrook and Springer Verlag, New York, pp. 313–345, 1998.

[FRE 08] FRÉON P., BOUCHON M., MULLON C., *et al.*, “Interdecadal variability of anchoveta abundance and overcapacity of the fishery in Peru”, *Progress in Oceanography*, vol. 79, pp. 401–412, 2008.

[FRE 10] FRÉON P., BOUCHON M., DOMALAIN G., *et al.*, “Impacts of the Peruvian anchoveta supply chains: from wild fish in the water to protein on the plate”, *GLOBEC International Newsletter*, pp. 27–31, April 2010.

[FRE 14a] FRÉON P., AVADÍ A., MARÍN W., *et al.*, “Environmentally extended comparison table of large vs. small- and medium-scale fisheries: the case of the Peruvian anchoveta fleet”, *Canadian Journal of Fisheries and Aquatic Sciences,* vol. 71, pp. 1–16, 2014.

[FRE 14b] FRÉON P., AVADÍ A., VINATEA R.A., *et al.*, "Boundary setting in lifecycle inventory analyses of complex and plural means of production: the case of the Peruvian industrial anchoveta fleet", *International Journal of LifeCycle Assessment*, vol. 19, pp. 1068–1086, 2014.

[FRE 14c] FRÉON P., SUEIRO J.C., IRIARTE F., *et al.*, "Harvesting for food versus feed: a review of Peruvian fisheries in a global context", *Reviews Fish Biology and Fisheries*, vol. 24, no. 1, pp. 381–398, 2014.

[FRE 83] FREON P., WEBER J., "Djifére au Sénégal : la pêche artisanale en mutation dans un contexte industriel", *Revue des Travaux de l'Institut des Pêches Maritimes*, vol. 47, nos. 3–4, pp. 261–304, 1983.

[FRE 99] FRÉON P., MISUND O.A., *Dynamics of Pelagic Fish Distribution and Behaviour: Effects on Fisheries and Stock Assessment*, Blackwell Science, Oxford, 1999.

[FRO 00] FRONTIER S., LEPRETRE A., "Développements récents en théorie des écosystèmes", *Annales de l'Institut Océanographique*, vol. 74, pp. 43–82, 2000.

[FRO 08] FRONTIER S., *et al.*, *Ecosystèmes: structure, fonctionnement, évolution*, Dunod, Paris, 2008.

[FRO 99] FRONTIER S., *Les écosystèmes*, PUF, Paris, 1999.

[GAR 03] GARCIA S.M., ZERBI A., ALIAUME C., *et al.*, "The ecosystem approach to fisheries: issues, terminology, principles, institutional foundations, implementation and outlook", *FAO Fisheries Technical Paper*, vol. 443, Rome, 2003.

[GAR 12] GARCIA S.M., KOLDING J., RICE J., *et al.*, "Reconsidering the consequences of selective fisheries", *Science*, vol. 335, pp. 1045–1047, 2012.

[GEO 71] GEORGESCU-ROEGEN N., *The Entropy Law and Economic Process*, Harvard University Press, Cambridge, 1971.

[GER 12] GERBER J.F., STEPPACHER R. (eds.), *Towards an Integrated Paradigm in Heterodox Economics: Alternative Approachs to the Current Eco-Social Crises*, Palgrave Macmillan, Basingstoke, 2012.

[GIA 07] GIAMPIETRO M., "Studying the "addiction to oil" of developed societies using the multi-scaled integrated analysis of social metabolism (MSIASM)", in BARBIR F., ULGIATI S. (eds.), *Sustainable Energy Production and consumption, Proceedings of NATO Workshop*, Springer, Amsterdam, 2007.

[GIA 98] GIAMPIETRO M., "Energy budget and demographic changes in socioeconomic systems", in DWYER S., GANSLOßER U., O'CONNOR M. (eds.), *Life Science Dimensions: Ecological Economics and Sustainable Use*, Filander Verlag, Fürth, 1998.

[GOR 54] GORDON H.S., "The economic theory of a common-property resource: the fishery", *Journal of Political Economy*, vol. 62, pp. 124–142, 1954.

[GOW 94] GOWDY J., *Coevolutionary Economics: The Economy, Society and the Environment*, Kluwer Academic, Boston, 1994.

[GRA 10] GRAEDEL T.E., ALLENBY B.R., *Industrial Ecology and Sustainable Engineering*, Pearson Education, Prentice Hall, 2010.

[GRA 35] GRAHAM M., "Modern theory of exploiting a fishery, and application to North Sea trawling", *Journal du Conseil, Conseil International pour l'Exploration de la Mer*, vol. 10, pp. 264–274, 1935.

[GRE 08] GRÉMILLET D., *et al.*, "A junk-food hypothesis for gannets feeding on fishery waste", *Proceeding of the Royal Society B*, vol. 275, pp. 1149–1156, 2008.

[GUI 08] GUILLOTREAU P. (ed.), *Mare economicum: enjeux et avenir de la France maritime et littorale*, Presses Universitaires de Rennes, Rennes, 2008.

[GUN 02] GUNDERSON L.H., HOLLING C.S. (eds.), *Panarchy: Understanding Transformations in Human and Natural Systems*, Island Press, Washington, 2002.

[GUN 98] GUNNINGHAM N., GRABOSKY P., *Smart Regulation: Designing Environmental Policy*, Clarendon Press, Oxford, 1998.

[HAR 68] HARDIN G., "The tragedy of the commons", *Science*, vol. 162, pp. 1243–1248, 1968.

[HEN 06] HENDRICKSON C.T., LAVE L.B., MATTHEWS H.S., *Environmental LifeCycle Assessment of Goods and Services: An Input-Output Approach*, Resources for the Future Press, Washington, 2006.

[HEN 12] HENRIKSSON P.J.G., GUINÉE J.B., KLEIJN R., *et al.*, "Lifecycle assessment of aquaculture systems – a review of methodologies", *International Journal of LifeCycle Assessment*, vol. 17, pp. 304–313, 2012.

[HJO 14] HJORT J., "Fluctuations in the great fisheries of northern Europe viewed in the light of biological research", *Rapports et Procés-Verbaux des Réunions du Conseil International pour l'Exploration de la Mer*, vol. 19, pp.1–228, 1914.

[HOL 86] HOLLING C.S., "The resilience of terrestrial ecosytems: local surprise and global change", in CLARCK W.C., MUNN R.E. (eds.), *Sustainable Development of the Biosphere*, Cambridge University Press, Cambridge, 1986.

[HUP 04] HUPPES G., *et al.*, Lifecycle costing and the Environment, CLM Report, 2004.

[HUX 84] HUXLEY T.H., "The fisheries exhibition", *Literature*, vol. 4, pp. 1–22, 1884.

[IFR 07] IFREMER, L'approche écosystémique des pêches : quelles priorités pour la recherche ?, IFREMER Report , May 2007.

[ISO 06] ISO, ISO 14040, Environmental management, Lifecycle assessment: Principles and Framework, International Organization for Standardization, 2006.

[ISO 14a] ISO, "ISO 14000 – Enviornmental management", http://www.iso.org/iso/iso14000, 2014.

[ISO 14b] ISO, "ISO 26000 – Social responsibility", http://www.iso.org/iso/iso/home/standards/iso26000.htm, 2014.

[JOL 10] JOLLIET O., SAADE M., CRETTAZ P., SHAKED S., *Analyse du cycle de vie : comprendre et réaliser un écobilan*, 2nd ed., Presses Polytechniques et Universitaires Romandes, Lausanne, 2010.

[JOR 00] JØRGENSEN S.E. (ed.), *Thermodynamics and Ecological Modeling*, Lewis Publishers, CRC Press, Boca Raton, 2000.

[JOR 07] JORGENSEN S.E., FATH B.D., *et al.* (eds.), *A New Ecology: Systems Perspective*, Elsevier Academic Press, Amsterdam, 2007.

[JOR 09] JØRGENSEN S.E. (ed.), *Ecosystem Ecology*, Elsevier Academic Press, Amsterdam, 2009.

[JOR 10] JØRGENSEN S.E., FATH B.D. (eds.), *Global Ecology: A Derivative of Encyclopedia of Ecology*, Elsevier Academic Press, Amsterdam, 2010.

[KLO 12] KLÖPFFER W., "The critical review of lifecycle assessment studies according to ISO 14040 and 14044", *International Journal of LifeCycle Assessment*, vol. 17, no. 9, pp. 1087–1093, 2012.

[KOO 10] KOOIJMAN S.A.L.M., *Dynamic Energy Budget Theory in Metabolic Organisations*, 3rd ed., Cambridge University Press, Cambridge, 2010.

[KRO 07] KRONENBERG J., *Ecological Economics and Industrial Ecology: A Case Study of the Integrated Product Policy of the European Union*, Routledge, London, 2007.

[LAN 14a] LANGLOIS J., FRÉON P., STEYER J.P., *et al.*, "Sea use impact category in lifecycle assessment: state of the art and perspectives", *International Journal of LifeCycle Assessment*, vol. 19, pp. 994–1006, 2014.

[LAN 14b] LANGLOIS J., FRÉON P., STEYER J.P., *et al.*, "New methods for impact assessment of biotic-resource depletion in LCA of fishery: theory and application", *Journal of Cleaner Production*, vol. 73, pp. 63–71, 2014.

[LAN 14c] LANGLOIS J., FRÉON P., STEYER J.P., *et al.*, "Sea use impact category in lifecycle assessment: characterization factors considering life support functions", *International Journal of LifeCycle Assessment*, 2014.

[LEF 99] LE FUR J., CURY P., LALOË F., *et al.*, "Co-viabilité des systèmes halieutiques", *Nature, Sciences, Sociétés*, vol. 7, no. 2, pp. 19–32, 1999.

[MAR 68] MARGALEF R., *Perspectives in Ecological Theory*, University of Chicago Press, Chicago, 1968.

[MAR 05] MARGALEF R., *Ecologìa*, Omega, Barcelona, 2005.

[MAR 07] MARTINET V., THÉBAUD O., DOYEN L., "Defining viable recovery paths toward sustainable fisheries", *Ecological Economics*, vol. 64, pp. 411–422, 2007.

[MAR 91] MARÉCHAL J.P., *Le prix du risque: l'économie au défi de l'environnement*, CNRS Press, Paris, 1991.

[MCG 01] McGLADE J.M., "Governance and sustainable fisheries", in VON BODUNGEN B., TURNER R.K. (eds.), *Science and Integrated Coastal Management*, Dalhem University Press, Dalhem, 2001.

[MEA 72] MEADOWS D., MEADOWS D.L., RANDERS J., *et al.*, *The Limits to Growth*, Universe Books, New York, 1972.

[MES 08] MESNIL B., "Public-aided crises in the French fishing sector", *Ocean & Coastal Management*, vol. 51, pp. 689–700, 2008.

[MON 14a] MONACO A., PROUZET P. (eds.), *Ecosystem Sustainability and global change*, ISTE, London and John Wiley & Sons, New York, 2014.

[MON 14b] MONACO A., PROUZET P. (eds.), *Development of Marine Resources*, ISTE, London and John Wiley & Sons, New York, 2014.

[MOR 80] MORIN E., *La Méthode, tome II: la Vie de la Vie*, Le Seuil, Paris, 1980.

[MUL 05] MULLON C., FRÉON P., CURY P., "Dynamics of collapse in world fisheries", *Fish and Fisheries*, vol. 6, no. 2, pp. 111–120, 2005.

[NES 07] NESS B., URBEL-PIIRSALU E., ANDERBERG S., *et al.*, "Categorising tools for sustainability assessment", *Ecological Economics*, vol. 60, pp. 498–508, 2007.

[NEW 04] NEWELL R.G., *Maximising Value in Multi-species Fisheries*, New Zealand Fulbright Program, Wellington, New Zealand, 2004.

[NIC 77] NICOLIS G., PRIGOGINE I., *Self-Organization in Non-équilibrium Systems: From Dissipative Structures to Order Through Fluctuations*, John Wiley & Sons, New York, 1977.

[ODU 07] ODUM H.T., *Environment, Power and Society for the Twenty-first Century: The Hierarchy of Energy*, Columbia University Press, New York, 2007.

[ODU 71] ODUM E.P., *Fundamentals of Ecology*, 2nd ed., Saunders, Philadelphia, 1971.

[OEC 07a] OECD-FAO, Agricultural Outlook 2007-2016, OECD and FAO, 2007.

[OEC 07b] OECD, "Globalization and fisheries", *Proceedings of an OECD-FAO workshop*, OECD, 2007.

[ORO 96] O'ROURKE D., CONNELLY L., KOSHLAND C.P., "Industrial ecology: a critical review", *International Journal of Environment Pollution*, vol. 6, no. 2/3, pp. 89–112, 1996.

[PAS 10] PASSET R., *Les grandes représentations du monde et de l'économie à travers l'Histoire: de l'univers magique au tourbillon créateur*, Les liens qui libèrent, Paris, 2010.

[PAS 79] PASSET R., *L'économique et le vivant*, Petite Bibliothèque Payot, Paris, 1979.

[PAU 02] PAULY D., CHRISTENSEN V., GUENETTE S., *et al.*, "Towards sustainability in world fisheries", *Nature*, vol. 418, pp. 689–695, 2002.

[PEL 07] PELLETIER N.L., AYER N.W., TYEDMERS P.H., *et al.*, "Impact categories for LifeCycle Assessment Research of seafood production systems: review and prospectus", *International Journal of LifeCycle Assessment*, vol. 12, no. 6, pp. 414–421, 2007.

[PER 96] PERMAN R., MA Y., MCGILVRAY J., *Natural Resources & Environmental Economics*, Longman, London, 1996.

[PIM 84] PIMENTEL D., HALL C.W. (eds.), *Food and Energy Resources*, Academic Press, New York, 1984.

[PIT 03] PITCHER T.J., CHEUNG W.W.L., "Fisheries: hope or despair?", *Marine Pollution Bulletin*, vol. 74, no. 2, pp. 506–516, 2003.

[PRE 08] PRÉ CONSULTANTS, SimaPro Database Manual: Methods library, PRé Consultants, 2008.

[RAM 02] RAMADE F., *Eléments d'Ecologie: écologie appliquée*, Dunod, Paris, 2002.

[RII 08] RIISGAARD L., BOLWIG S., MATOSE F., *et al.*, "A strategic framework and toolbox for action research with small producers in value chains", *DIIS Working Paper*, no. 17, 2008.

[ROB 02] ROBÈRT K.H., SCHMIDT-BLEEK B., ALOISI DE LARDEREL J., *et al.*, "Strategic sustainable development: selection, design and synergies of applied tools", *Journal of Cleaner Production*, vol. 10, pp. 197–214, 2002.

[SCH 57] SCHAEFER M.B., "A study of the dynamics of the fishery for yellowfin tuna in the Eastem Tropical Pacific Ocean", *Inter-American Tropical Tuna Commission Bulletin*, vol. 2, pp. 247–285, 1957.

[SCH 94] SCHNEIDER E.D., KAY J., "Life is a manifestation of the second law of thermodynamics", *Mathematical Computing Modeling*, vol. 19, pp. 25–48, 1994.

[SEU 08] SEURING S., MÜLLER M., "From a literature review to a conceptual framework for sustainable supply chain management", *Journal of Cleaner Production*, no. 16, pp. 1699–1710, 2008.

[SHI 10] SHIN Y.J., SHANNON L.J., BUNDY A. *et al.*, "Using indicators for evaluating, comparing, and communicating the ecological status of exploited marine ecosystems. 2. Setting the scene", *ICES Journal of Marine Science*, no. 67, pp. 692–716, 2010.

[SMI 08] SMIL V., *Energy in Nature and Society*, MIT Press, Cambridge, 2008.

[SOC 94] SOCOLOW R., ANDREWS C., BERKHOUT F., *et al.* (eds.), *Industrial Ecology and Global Change*, Cambridge University Press, Cambridge, 1994.

[SOT 08] SOTO D., AGUILAR-MANJARREZ J., HISHAMUNDA N. (eds.), "Building an ecosystem approach to aquaculture", *FAO Fisheries and Aquaculture Proceedings*, vol. 14, 2008.

[SUM 08] SUMAILA U.R., TEH L., WATSON R., *et al.*, "Fuel price increase, subsidies, overcapacity, and resource sustainability", *ICES Journal of Marine Science*, vol. 65, pp. 832–840, 2008.

[TAM 08] TAM J., TAYLOR M.H., BLASKOVIC V., *et al.*, "Trophic modeling of the Northern Humboldt Current Ecosystem, Part I: Comparing trophic linkages under La Niña and El Niño conditions", *Progress In Oceanography*, vol. 79, nos. 2–4, pp. 352–365, 2008.

[THE 98] THÉBAUD O., "Searching for common ground: economics, ecology and the management of martine living resources", in DWYER S., GANSLOßER U., O'CONNOR M. (eds.), *Life Science Dimensions: Ecological Economics and Sustainable Use*, Filander Verlag, Fürth, 1998.

[THO 80] THOMSON D., "Conflict within the fishing industry", *ICLARM Newsletter*, vol. 3, no. 3, pp. 3–4, 1980.

[TIE 00] TIETENBERG T., *Environmental and Natural Resource Economics*, Addison Wesley Longman, Reading, 2000.

[TIE 05] TIETZE U., THIELE W., LASCH R., *et al.*, "Economic performance and fishing efficiency of marine capture fisheries", *FAO Fisheries Technical Paper*, no. 48, 2005.

[TVE 05] TVETERÅS R., "The effect of regulatory regimes on productivity development in fisheries: a comparative country study", *3rd Biennial North American Association of Fisheries Economics Forum*, Vancouver, 25–27 May 2005.

[TVE 06] TVETERÅS R., KVALØY O., "Changes in the organization of value chains for food – the response from seafood sectors", in ASCHE F. (ed.), *Primary industries facing global markets: The supply chains and markets for Norwegian food*, chapter 3, Aschehoug, Oslo, 2006.

[TYE 04] TYEDMERS P., "Fisheries and Energy Use", in CLEVELAND C. (ed.), *Encyclopedia of Energy*, vol. 2, pp. 683–693, Elsevier, Amsterdam, 2004.

[TYE 05] TYEDMERS P., WATSON R., PAULY D., "Fueling global fishing fleets", *Ambio*, vol. 34, no. 8, pp. 635–638, 2005.

[ULA 00] ULANOWICZ R.E., *Growth and Development: Ecosystems Phenomenology*, Excel Press, San Jose, 2000.

[VAL 07] VALDIMARSSON G., "Fish in the global food supply chain", diapositives pour le *World Seafood Congress*, Dublin, Irlande, September 2007, Available at: www.globefish.org/files/.

[VAL 11] VALDIVIA S., UGAYA C., SONNEMANN G., *et al.*, (eds.), *Towards a LifeCycle Sustainability Assessment: Making Informed Choices on Products*, UNEP/SETAC LifeCycle Initiative, 2011.

[VAZ 12] VÁZQUEZ-ROWE I., HOSPIDO A., MOREIRA M.T., *et al.*, "Best practices in lifecycle assessment implementation in fisheries. Improving and broadening environmental assessment for seafood production systems", *Trends in Food Science and Technology*, vol. 28, no. 2, pp. 116–131, 2012.

[WAC 96] WACKERNAGEL M., REES W., *Our Ecological Footprint: Reducing Human Impact on the Earth*, New Society Publishers, Gabriola Island, 1996.

[WAR 99] WARNIER J.P., *Construire la culture matérielle : l'homme qui pensait avec ses doigts*, PUF, Paris, 1999.

[WCE 87] WCED (WORLD COMMISSION ON ENVIRONMENT AND DEVELOPMENT), *Our common Future*, Oxford University Press, Oxford, 1987.

[WEI 09] WEIZSÄCKER VON E., HARGROVES K., SMITH M.H. (eds.), *Factor Five: Transforming the Global Economy Through 80% Improvements in Resource Productivity*, Earthscan, London, 2009.

[WOR 08] WORDWATCH INSTITUTE, *Farming Fish for the Future*, Worldwatch Institute Edition, Washington, 2008.

[WOR 09] WORM B., HILBORN R., BAUM J.K., *et al.*, "Rebuilding global fisheries", *Science*, vol. 325, pp. 578–585, 2009.

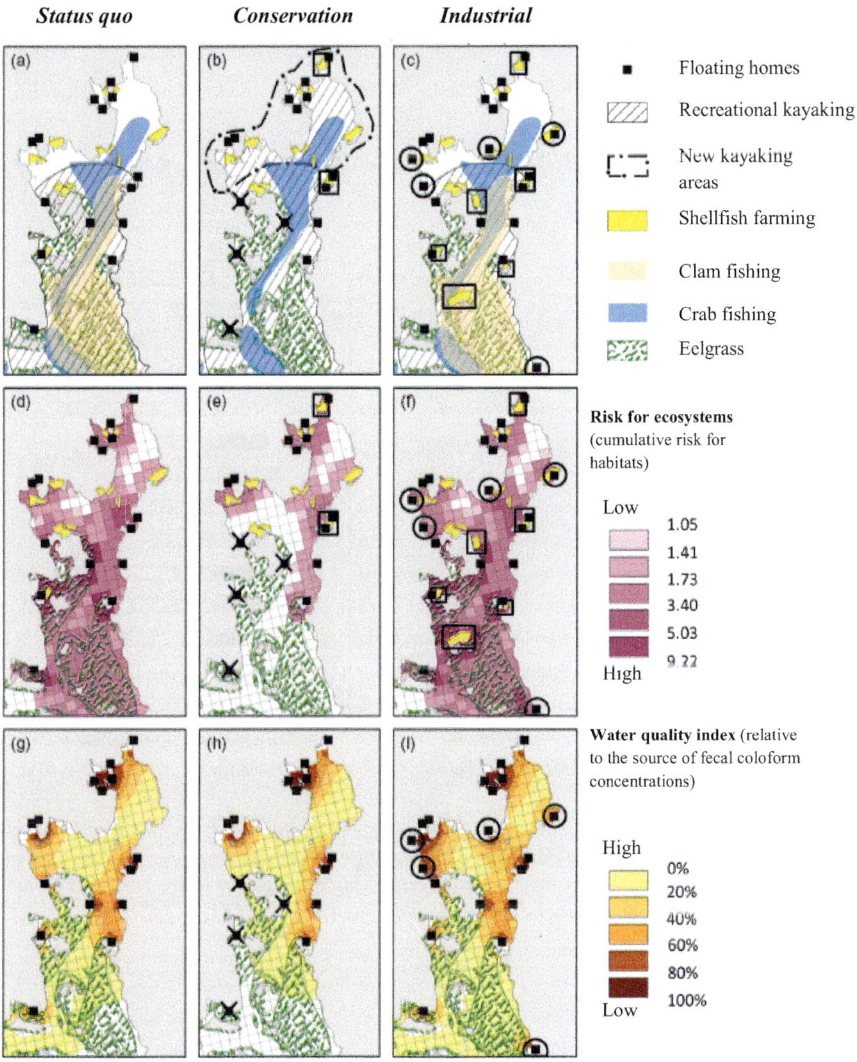

Figure 1.7. *Three management scenarios for Lemmens Inlet (a–c) and some InVEST (d–i) exit strategies [GUE 12]*

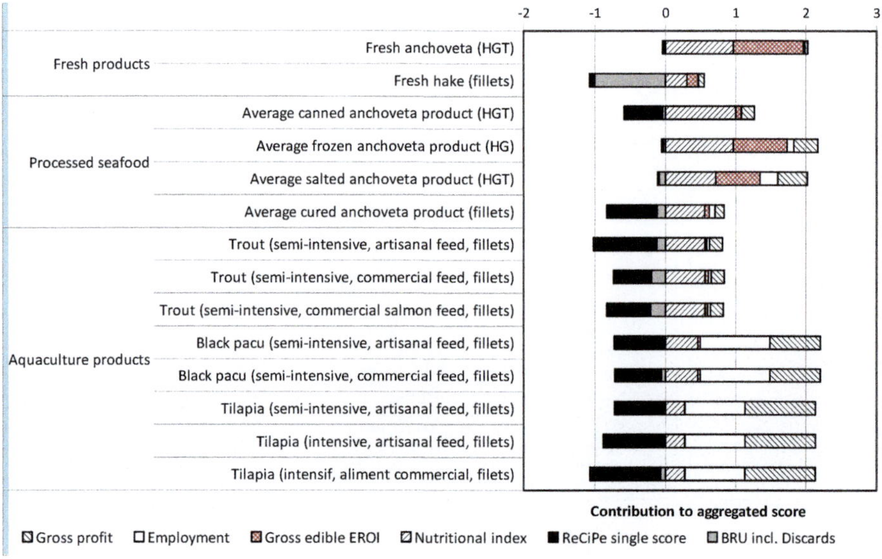

Figure 2.10. *Classification of all the DHC products from the supply chains of anchovy studied rendered at the doors of the factory, according to the set of proposed indicators, per ton of fish in the product (the shortest negative bars and the longest positive bars represent a better performance). HGT = headed, gutted, tailed [AVA 14d]*

Figure 2.11. *Semi-intensive aquaculture basin of black pacu (Colossoma macropomum) in the Amazon region of Iquitos, Peru. The average size of the basins is 0.86 ha, the production in the order of 10 t/ha/year, the production cycle of 12 months for an individual of average weight of 1.2 kg and the employment of 1 person/t (photo Angel Avadi)*

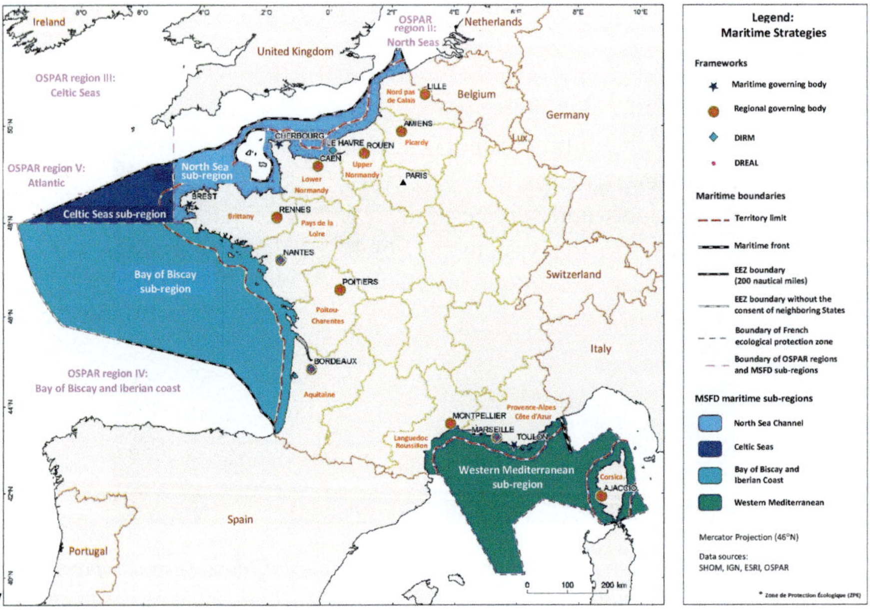

Figure 4.1. *Delimitation of submarine regions concerning France's seafronts (Source: Agency of marine protected areas, Ifremer)*

3

Fisheries Economics

3.1. Outline of fisheries economics

3.1.1. *Object of fisheries economics*

If we choose to retain a substantive definition for fisheries economics, following that of the economics proposed by Lionel Robbins[1], it can be said that it aims at studying the individual and collective choices in the alternative uses of scarce resources. These resources are here labor, economic capital (boats, fishing gears and on land facilities) and the natural capital (constituted of exploited stocks) in the pursuit of individual and collective objectives. It is important here to bring up the significant points of this apparently very simple definition. The notion of scarcity is central, the economy is only interested in the useful resources whose availability is finite, faced with unlimited individual and collective necessities, and which can be the subject of concurrent uses. Another important point is that economics is not concerned with the social opportunity of the objectives pursued in the use of resources. It can be considered that the latter relate to politics in the case of the collective choices or free will at the individual level; each individual is characterized by a utility function

Chapter written by Christian CHABOUD.

1 Economics is the science studying human behavior as a relationship between ends and rare means with alternative uses [ROB 47, p. 30].

that enables him or her to measure the relative attractiveness of the different possible choices in the use of resources available to him or her.

3.1.2. *Specific features and characteristics of fisheries economics*

3.1.2.1. *The renewable character of fisheries resources*

What are the specific features of fisheries economics within economics science? First, a particular characteristic of the exploited natural capital will be presented: its renewable character that consists of its endogenous capacity of reproduction. Fisheries economics is thus a subfield of natural resources economics within which a distinction is effected between the non-renewable resources, whose exploitation results inexorably in exhaustion (but at an optimal rhythm that should be determined in relation to the economic context, as proposed by the well-known Hotelling's model [HOT 31]), and those that are subject to natural processes of growth and mortality, independently of the uses that people make of it. This characteristic, which is found for other resources such as forests, for example, led us to address the optimal use of these resources according to the capital economics approach suggested by neoclassical bioeconomists [CLA 85, CLA 90, CLA 06, CON 10, HAN 93, GRA 04, GRA 06]. Within the renewable resources, a second distinction can be made [CLA 87] between those, qualified as exhaustible, whose natural productivity is affected by the intensity of their exploitation by humans (such as forests, pastures, fish stocks and other wild animal populations) and those whose productivity is not affected by the uses that are made of them (such as geothermal energy, solar energy and non-fossil water resources).

3.1.2.2. *Access and appropriation regimes of fisheries resources*

A second important characteristic of fisheries resources, from the economic point of view, is related to their access and appropriation regime. In economics, goods are generally classified according to two criteria: rivalry (when the use of certain goods by an economic agent has a negative effect on the quantity of these goods available to other individuals) and exclusivity (when not all individuals have access to these goods). It is possible to classify the different types of goods according to these two criteria. Most often, fisheries resources are a common property (still called impure public goods) for which there is rivalry, but pure non-exclusivity or limited exclusivity[2] between users, which differ from private goods (totally exclusive and rival) and pure public goods (non-rival and non-exclusive (such as the quality of the air and the landscape)).

Historically, the general regime of common goods of fish stocks has not always been the norm [SCO 07]. This is the outcome of the legal controversy between the Dutch jurist Hugo Grotius [GRO 16] and the British in the 17th Century, favorable to the Dutch jurist, who founded the general principle of freedom of navigation and access to marine resources. The character of the common resource of fish stocks is at the origin of many works, which, as a result of the seminal articles of Scott Gordon [GOR 54] and Garrett Hardin [HAR 68], consider that the common access to renewable resources leads inexorably to their overexploitation. Hardin starts from the example of communal pastures, such as existed in the past in Europe, and shows that each rancher benefits from increasing the size of his herd because the profit that he gains belongs integrally to him while the costs resulting from the loss in forage resource are shared between all cattlemen. This phenomenon, which relies on the incompatibility between individual rationality and collective interest, has even been formalized in game theory by John Nash with the famous "prisoner's dilemma" [BIN 99]. Hardin's example can be generalized to the entirety of the common resources: wildlife, forests and of course fish stocks. From his conclusions, a number of proposals are derived to resolve the "tragedy of the commons" in two main directions. First, the implementation of public policies (regulations) aimed at internalizing the costs, that is to say making the economic agents bear the individual consequences of their decisions, their economic calculation

2 When it is exercised between individuals, pure exclusivity will be used; otherwise, it will be limited exclusivity when it is exercised between different groups of individuals but not between individuals within these groups.

then being no longer distorted by the existence of negative externalities[3]. The second direction aims at promoting property rights, if possible transferable, in order to reduce the inefficiency in the exploitation of common resources [ARN 07]. As will be seen further in this chapter, this orientation has led to the instauration of individual transferable quotas (ITQ) in many fisheries. The ongoing reform of the common fisheries policy (CFP) in Europe intends to give an important role to this management tool, which has caused a large number of debates.

However, not all authors agree with those who consider that the status of common resource is an original sin which can only lead to their misuse from a collective point of view. Research in economic anthropology and institutional economics has fuelled the thought of the "school of the commons" with Elinor Ostrom [OST 90, OST 10] as a leader, whose works on the subject were rewarded by the Nobel Prize in Economics in 2009. Unlike Gordon, Hardin and their successors, the supporters of the "school of the commons", consider that there exist numerous examples of collective actions allowing the efficient use of common natural resources. Nonetheless, they recognize that collective action can fail to achieve efficiency and conditions must be fulfilled in such a manner that it will lead to an optimal use of the resources from a collective perspective.

Among these are clearly defined limits of the resources and of the community of legitimate users (and therefore the exclusion rules), the rules tailored to the objectives of the community of users and to local conditions, the participation of individuals in the definition and amendment of the rules, a governance mode imposing on users to give accounts to the whole community, a system of progressive sanctions against users not respectful of the rules and an inexpensive mode of conflict resolution. In the particular area of fisheries, history has shown

3 Externality (or external effect) is used to denote the situations where the decisions of one economic agent affect the well-being of one or more other agents without any existing compensatory transactions between them. Negative externality will be used when well-being is reduced, as in the case of industrial pollution which affects the health of the other agents, or also in the case of fisheries where the extraction of resource by an individual reduces the quantity available to others. Conversely, positive reciprocal externalities are observed when, for example, arboriculture and apiculture coexist in the same space.

that the traditional management systems of fisheries have fulfilled these requirements. Some examples include the French *prud'homies* and the Spanish *cofradias* which have existed since the Middle Ages and continue to exercise the responsibilities of local management alongside more modern instances such as the regional committees of fisheries in France. Community management modes have given rise to an extensive literature which often concerns the Southern countries [BER 1], notably the insular fisheries of the Pacific [RUD 84].

If at the theoretical level, the positions of the two previous schools always tend to be very contrasting and to inspire heated debate, it is important to recognize that the two approaches have merits that should be pointed out depending on contexts. To this effect, a very useful and relevant distinction between local common resources and global common resources has been proposed by Dasgupta [DAS 93]. The former are those that can be managed at the community or local level (e.g. reef fish, some shellfish or coastal lagoons fisheries, lake fisheries, etc.); the latter are those whose characteristics require that they can only be managed at scales exceeding those of the regulating capacity of communities or local management instances (cases of migratory fish such as tuna, coastal species whose stocks cover large surfaces or even demersal fish such as cod in the North Atlantic, etc.). In this case, it seems difficult that an effective collective action could exist between users who do not know each other and do not share common objectives; their coordination is governed by the action of states or of international institutions (regional fishery organization).

3.1.2.3. *The issue of risk and uncertainty in fisheries and its consequences*

If the classic distinction by F. Knight[4] [KNI 21] between risk and uncertainty is used, it is obvious that the two types of situation, risk and uncertainty, are observed in fisheries [CHA 95].

4 An environment or universe Ω is said to be risky if for each of the states e_i can be associated with a probability p_i such that (discrete formulation): $\sum_i p_i = 1 \quad 0 \le p_i \le \forall i$. The term "uncertainty" will be used when the different events or states cannot be probabilized.

The producers are mainly vulnerable to the risk linked to the resource. The latter is subject to unpredictable natural variations (short-term fluctuations in the availability and medium- and long-term changes of the abundance), but especially to unpredictable decisions from other producers which affect the results of all the fishermen.

The second source of uncertainty concerns the fish market. According to the configuration of fishery systems (isolated fishery selling on local market, fishery with several landing centers selling on the same major consumption centers, or world market-oriented fishery), variability in prices will be affected by fluctuations in the local supply or by total landings or even by factors totally exogenous to the local or national economy (supplies of concurrent fisheries for the same products, markets for substitutable products, protectionist practices, etc.). The uncertainty of the input market is also important, particularly with regard to fuel. Another major source of uncertainty is constituted by "perils of the sea": losses of equipment or vessels and accidents at work. Even in countries where safety regulation is highly enforced, it is considered that the fisheries sector is the most exposed to accident at work risk [AND 83]. The situation is obviously more serious in Southern countries where regulations are laxer or virtually non-existent in the case of artisanal fishery (cases, for example, of canoe fishery in Senegal [THI 12] or the traditional fishery in Madagascar [CAV 08]).

Another category of risk and uncertainty is linked to the availability and quality of production factors, other than biological resources, such as labor and technical capital (boats, fishing gear, etc.). For example, with regard to artisanal fisheries, it is often noticed that formal labor contracts in which the commitments of both parties are clearly defined do not exist. The most qualified crew members can thus easily move between fishing units and the owners are faced with the risk of stopping or decreasing their activity if they are not able to guarantee the stability in crew composition. The quality of technical equipment at sea is crucial, because they are used in a complementary way and the failure of any one of them undermines the effectiveness of the whole. The motorization of canoes in West Africa, often presented as a technological and institutional success story [CHA 91], has made artisanal fishery not only dependent on the variability of the price of fuel, but also on the reliability of outboard engines and the availability of spare parts.

The institutional risks and uncertainties[5] must finally be mentioned and they are far from negligible. They are of two kinds: those related to the microeconomic relationships between individual actors, and those that are sector-based or related to macroeconomics. Among the first are the practices of income diversion at the expense of ship-owners or fishermen. Among these is the possibility for the fishermen to sell a part of the catch at the expense of the ship-owner or, conversely, the possibility for the ship-owner to under-declare the turnover of the fishing unit and thus reduce the crew share. It also concerns the non-respect of trade commitments between intermediaries and fishermen, including the renegotiation of the initial ex-vessel prices (traders benefiting from an asymmetry of information on the fish market). Conversely, fishermen may not respect the commitments of exclusive sale with a trader or even worse the reimbursement of monetary loans granted by the latter (in Southern countries, it is common for commercial intermediaries to grant credits to their supplier in order to gain their loyalty or enable them to cope with unforeseen events [CHA 89, PLA 85]). The sectoral institutional risks, which can be characterized as external, relate in the first place to the growing importance of the administrative regulations; the permanent adjustments of the public fishing policies (also justified by the search for a more flexible management of the resources and activities) may, for example, impose changes in the characteristics of vessels, equipment and, therefore, in fishing strategies. The provisioning of the processing industries can become very irregular through the instauration of fishing seasons and of catch quotas which may encourage fishing and processing overcapacity (case of cephalopod fisheries in Northwestern Africa managed by means of

5 The terms "institutions" and "institutional" are considered here with a double meaning: first, the one used by the economics science of arrangements, conventions, norms and rules that govern the behavior of economic agents; then that of the organizations (among other things, public) in charge of the implementation of public policies (for example, the Fisheries Committee of the European Union, the supervision of marine fisheries and aquaculture at the national level, etc.).

"biological rest"). The increasing role of supranational institutions is itself a risk factor for fishermen and ship-owners because the bargaining power of professional lobbies is less than it is in the national framework.

What are the consequences of the importance of risk and uncertainty in fisheries? This encourages first resorting to forms of organization and insurance mechanisms that will allow sharing risks between economic actors. The near non-existence of fixed wages in fishery and the persistence of the shared-wage system (similar to the contract of sharecropping in agriculture) are a form of response to particular risks in fishery: strong irregularity of the turnover in the short term and the impossibility for the ship-owner, not being on board, to check the enthusiasm of the crew to work [SUT 79]. The informal agreements between fishermen and fishing interests can also be interpreted as a way to share resource and market risks [CHA 89, WIL 80]. The second important point is the impact on the behavior of the fishermen; the difficulty of planning in the medium and long terms for their activity gives them an incentive to adopt a myopic and opportunistic management[6], not really favorable to a prudent and durable business direction.

3.2. The bioeconomic approach of fisheries

Bioeconomics is the analysis of the behavior of the fishing system represented by the interactions between the different forms of capital (natural capital, economic capital), the labor force, the decisions (choice of fishermen with respect to investment, but also the public policies with regard to the development of the fishery) and, finally, elements that are exogenous but whose influence is very significant (inputs or fish products market, national or supranational legislative and regulatory framework, etc.). As its name indicates, "bioeconomics" refers at the same time to both economics and

6 During an interview with the manager of an industrial fishing firm from Dakar in 1990, I asked a question about the investment strategy of this firm (the question was simply "when and how do you decide on the purchase or renewal of a boat"); I was told that this was mainly dependent on the current cash flow of the company ... regardless of considerations on the evolution of the resource or of the market.

fisheries biology tools. The borrowing from biological sciences relates essentially to the standard population dynamics models which will be used to represent the evolution of the resource in response to the choice of exploitation. The term "bioeconomics" first appeared at the beginning of the 20th Century in the writings of the great Russian fishery scientist Baranoff, although he did not refer much to economic factors [GOR 54]. An extensive literature in bioeconomics has developed since the mid-1950s, marked by the production of the bioeconomic reference model called Gordon–Schaefer's model, from which many theoretical and methodological developments were made. We will therefore begin with its presentation and its results, and discuss them before moving on to other main models proposed as a result of this pioneering work.

3.2.1. *Gordon–Schaefer's model*

This reference model is derived from the connection of the parallel works of the economist Gordon [GOR 54] and the biologist Schaefer [SCH 54].

3.2.1.1. *The biological components of the model*

The exploited resource is composed of a monospecific stock whose biomass X is subject, in the absence of fishery, to the only influences of natural mortality and the arrival of new individuals. More specifically, the instantaneous variation of biomass is represented by following the equation:

$$\frac{\partial X}{\partial t} = B - D \qquad [3.1]$$

where B and D are, respectively, the instantaneous births and deaths. They are considered proportional to X: $B = b.X$ and $D = d.X$, where b and d are the instantaneous rates of birth and mortality. The intrinsic rate growth r is equal to the difference between the rates of birth b and mortality d, and the previous equation can now be written as:

$$\frac{\partial X}{\partial t} = r.X \qquad [3.2]$$

If equation [3.2] is integrated over time, biomass X follows an exponential growth of the form $X_t = X_0 e^{rt}$. It is obvious that such a growth is impossible in an ecosystem within which the resources essential for the renewal or the growth of a population are, by definition, finite. It is therefore considered that each stock may not exceed a value K (called carrying capacity) and that its growth will tend toward zero when it approximates K. A new expression of the growth (denoted as $G(X)$ further in the presentation) is thus obtained:

$$\frac{\partial X}{\partial t} = r.X.\left(1 - \frac{X}{K}\right) = G(X) \qquad [3.3]$$

Figures 3.1 and 3.2, respectively, represent the instantaneous growth as a function of the biomass and the evolution in time of the biomass from $X_0 = 0$, with $K = 100$ and $r = 0.3$. The growth is maximal for $X = K/2$, which corresponds to the inflection point of Figure 3.2, increasing for $X < K/2$ and decreasing beyond that.

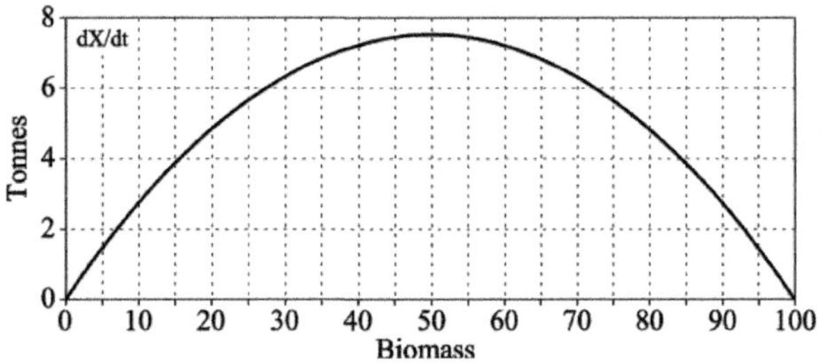

Figure 3.1. *Biomass growth for $X_0 = 0$, $K = 100$ and $r = 0.3$*

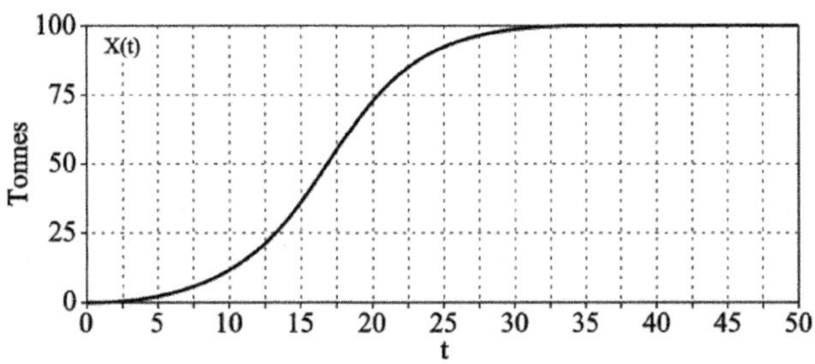

Figure 3.2. *Evolution in time of the biomass in the absence of exploitation*

Other forms of the growth function are possible. Clark [CLA 06], following Fox [FOX 70], proposed a more general form, the initial Gordon–Schaefer's model being obtained for $\alpha = 1$:

$$G(X) = r.X.(1 - X / K)^{\alpha} \qquad [3.4]$$

3.2.1.2. *Introduction of exploitation in the model*

Fishery consists of the extraction of a resource by humans, which will be a function of the biomass abundance (natural capital) and the human (labor) and material (economic capital) means that are allocated for this purpose. This production function is presented in equation [6.5], where Q is the instantaneous catch, C is the economic capital and L is the labor:

$$Q = Q(X, C, L) \qquad [3.5]$$

The theory of fisheries simplifies the previous formula by referring to the concept of fishing effort (E), defined as "a measure of all the means of capture applied by fishermen on a stock, during a given time interval" [LAU 81]. C and L are thus grouped in E. This simplification is not evident insofar as different combinations of C and L (qualified as technologies) can be used to produce the same level of fishing effort. To overcome this difficulty, a number of standardization procedures of fishing effort are used. Finally, the measurement of fishing effort will be adapted according to the type of fishing units considered: hours at sea or trawling for trawlers, number of hooks immersed in the case of longliners, length of nets with regard to netters, etc. The difficulty of this choice derives from the fact that economists and biologists will not necessarily make the same choice *a priori* during the measurement of the effort: the first will seek a measurement to relate to the fishing costs and to the catches while the second will likely only be interested in the relationship with the catches.

If we leave aside this difficulty, the production function can now be written as $Q = Q(X, E)$ and it is convenient to specify more explicitly its form. In this direction, a new concept is proposed by the theory of fisheries: catchability, which will be denoted as q. Formally, it is defined as the probability that a biomass unit is captured when an effort unit is applied to the stock, during a given time interval. In reality, this concept covers a much more complex reality, which is the encounter between the fisherman and the fish. On the fisherman side, catchability will be influenced by the human experience, by the technical efficiency of the equipment, by the existence or non-existence of cooperation between fishing units, etc. On the fish side, it will be dependent on its availability (presence in the fishing areas) and its vulnerability to the fishing gear when it encounters it. It can undoubtedly be considered that catchability is one of the most important concepts of fisheries science, which is still not yet fully described and understood. The hypotheses concerning catchability are particularly critical. In the model outlined here, it will be considered that the catchability q is independent of the biomass X. Such a hypothesis has to be reconsidered in the case of species with aggregative behavior [CLA 06], such as anchovies. The famous example of the collapse of the resource of the Peruvian anchoveta in 1972–1973 showed that in spite of the decline in abundance preceding this event, catches have not diminished proportionally to the biomass due to concentrating behavior of the resources of the coast [THO 81], which signifies an increase in catchability whereas biomass has decreased. These limits being established, the production function, based on the formal definition of catchability proposed earlier, then becomes:

$$Q = q.E.X \qquad [3.6]$$

By now introducing the extraction effected by fishery, a new equation of the instantaneous biomass variation is obtained, which is equal to the difference between the natural growth $G(X)$ and the catches:

$$\frac{\partial X}{\partial t} = r.X.\left(1 - \frac{X}{K}\right) - q.E.X \qquad [3.7]$$

The exploitation will be considered in equilibrium if the biomass remains constant; hence, $\frac{\partial X}{\partial t} = 0$ and therefore $G(X) = q.E.X$. For each level of fishing effort $E_1, E_2, \ldots, E_n$, there is then a level of biomass $X_1, X_2, \ldots, X_n$ such that the previous condition is verified. The determination of the catch equilibrium for four levels of fishing effort (100, 200, ..., 400) is shown in Figure 3.3. These equilibriums are determined by the intersection of the growth curve $G(X)$ and the straight lines of the catches corresponding to each level of fishing effort, and whose slope is the product of the effort by the catchability. These equilibriums are stable: in their neighborhood on the right, $G(X) < q.E.X$; therefore, X decreases to return to the initial equilibrium (the inverse being checked on the left).

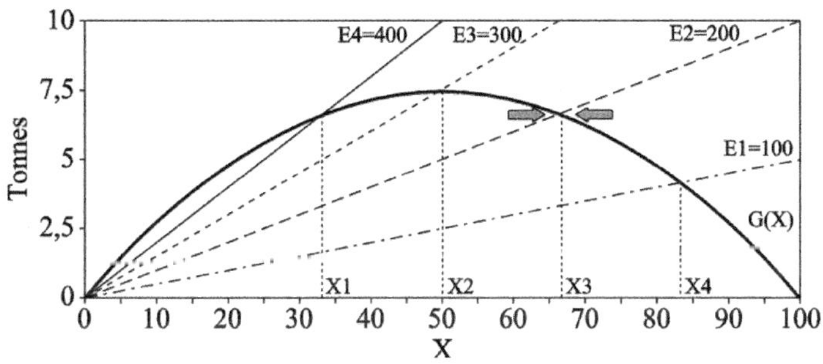

Figure 3.3. *Determination of the equilibrium catch*

In this example, it can be observed that the increased effort from 0 to 300 units is accompanied by an increase in catches, but any increase beyond 300 results in a reduction of the equilibrium catch. We will give more details about this point later. The previous developments can be deducted from two equilibrium relationships: the first between the biomass and the fishing effort (equation [3.8]), and the second between the fishing effort and catches (equation [3.9]):

$$\frac{\partial X}{\partial t} = 0 \Rightarrow r.X.\left(1 - \frac{X}{K}\right) = q.E.X \Rightarrow X = K.\left(1 - \frac{qE}{r}\right) \qquad [3.8]$$

The equilibrium biomass is a decreasing linear function of the effort, and this negative relationship will be much more significant as q increases and r decreases. By replacing X by its previous expression as a function of E, we can reformulate the production function (see equation [3.6]):

$$Q = q.E.K.\left(1 - \frac{qE}{r}\right) = q.E.K - \frac{q^2}{r}E^2K \qquad [3.9]$$

The equilibrium relationship between the equilibrium catch Q and the fishing effort E is then of a quadratic form. In calculating its first derivative with respect to E and by making it equal to zero, the value of the fishing effort is determined, which maximizes the equilibrium catch; hence, $E = r/2q$. By replacing E by $r/2q$ in equation [3.9], the maximum equilibrium catch or MSY[7] is obtained, which is equal to $rk/4$.

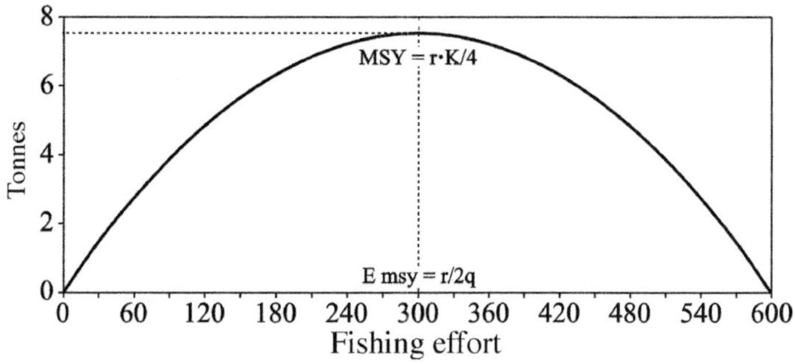

Figure 3.4. *Equilibrium relationship between fishing effort and catches*

Two important results from this model must be emphasized here. First, the resource is more productive when exploited. This counter-intuitive result is easily explained by the constraint of equilibrium and the underlying growth model: when the biomass is close to K, its

7 *Maximum Sustainable Yield.*

natural growth is very low and therefore also the equilibrium catch. The second result shows that even with very high levels of fishing effort, it is not possible to totally exhaust the resource because the equality between extraction and growth is always respected. This inability to explain the collapse of an exploited stock has resulted on the part of some authors to propose new model formulations reconsidering the shape of the growth function *G(X)*. An example proposed by Clark [CLA 90] is that of "critical depensation" models. In order to achieve this, it proposes to introduce a critical value of the biomass K_0, called minimum viable population level, below which the growth *G(X)* becomes negative (Figure 3.5).

In Figure 3.5, two equilibriums (*A* and *B)* appear between the growth *G(X)* and the production function *q.E.X* for a given level of effort. Equilibrium *B* is similar to that obtained in Figure 3.3; it is therefore stable. Equilibrium *A*, in contrast, is unstable. Move away to the right, then *G(X)* > *q.E.X* and biomass will stretch up to the value that corresponds to equilibrium *B*. If we move away from *A* toward its left, then *G(X)* < *q.E.X* and the biomass will decrease to reach its minimum viable value *Ko* and then will collapse. Such a situation could explain the collapse of the population of marine mammals, such as the blue whale in the Antarctic. For some authors [MAR 14], this could also explain the absence of recovery in populations that have decreased to very low levels of abundance in spite of catch moratorium, as for cod in the North Western Atlantic.

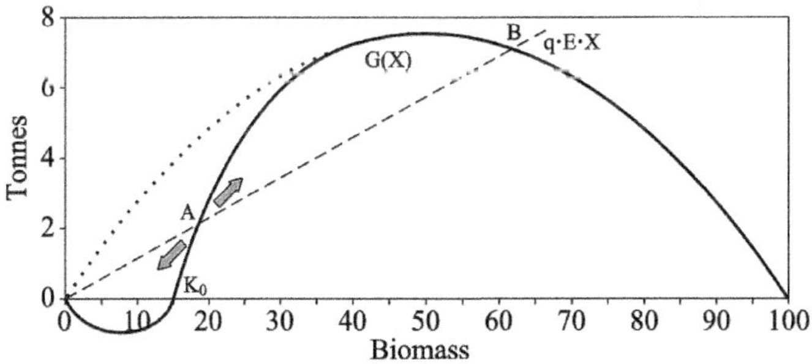

Figure 3.5. *Possibility of the resource collapse in the case of a critical depensation growth function*

3.2.1.3. *The economic components of the model*

To ensure that the model can claim the qualifier bioeconomic, it is appropriate from now on to introduce the three main economic elements that are the formation of prices (demand function), the formation of production costs (cost function) and the exploiter's behavior rules.

3.2.1.3.1. The demand function

The demand function $Q = Q(P)$ indicates the quantity Q of economic good that will be demanded for different values of the price P, by period of time, all things being equal (income, prices of other goods, preferences of the consumer, etc.). First, it is considered that a curve of individual demand exists for each good and each consumer. This curve has a negative slope, due to the decreasing marginal utility[8] of goods. The market demand for a good is constituted by the horizontal summation of all the individual requests. There will be references to the function of inverse demand $P = F(Q)$ when the prices are expressed as a function of the good quantities. The market demand function must therefore be interpreted as the function that indicates the price that can be obtained on the market for different production levels of enterprises. It allows us to deduct the income $P.Q$ that a sector of activity can expect, depending on the quantity produced Q. In the case of a linear inverse demand function of the form $P = a+b.Q$ with $b < 0$, as in Figure 3.6, the curve of income $P.Q$ is parabolic and its maximum is obtained for $Q = -a/2b$.

A very important distinction should be recalled here when the changes of some goods in the market are analyzed. The change of demanded quantity will be mentioned, when for a given demand function $Q = Q(P)$, a price change ΔP causes a variation in quantity Q. On the other hand, the change of demand (direct or reverse) will be

8 Utility represents the individual well-being obtained from the consumption of goods; it is subjective (namely specific to each individual). The marginal utility is the increase of utility allowed by the consumption of an additional unit of goods. It is considered as decreasing because the utility obtained from the last consumed unit is always less than the one permitted by the previous unit. As a consequence, a consumer will not accept to buy an additional unit of goods unless its price decreases.

mentioned when the demand function shifts depending on various factors.

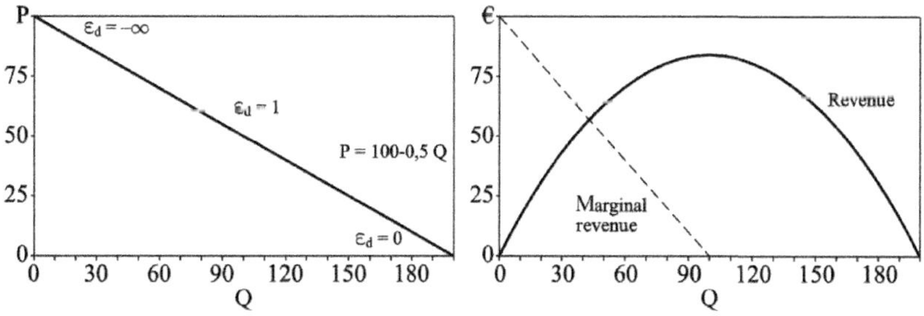

Figure 3.6. *Linear inverse market demand function: evolution of income and marginal revenue as a function of quantity*

Among various factors, the income of the consumer and the prices of other goods are the important ones. For the so-called ordinary goods (it can be assumed that this is the case for the majority of fish products), the demand function shifts toward the right side when the revenue increases. For the same price level, the consumer whose revenue has increased will consume a larger amount of the good being considered. In the case of inferior goods (which can be the case of some small pelagics such as the sardine or the sardinella), the inverse will be observed.

The effect of prices of other goods on the demand of a particular kind of goods will depend on their substitutability or complementarity relationship. Two types of goods are said to be substitutes if they allow the same necessity to be satisfied, without being perfectly homogeneous. If the price of one of the substitutable commodities increases, the demand of the other substitute decreases. Meat and carnivorous fish can be considered as substitutes on the markets of developed countries. Complements are commodities that must be consumed together to satisfy the same needs. In the industrial field, the cases of cement and iron for concrete or even cars and tires can be cited. On the food market, the cases of strict complementarity are more difficult to find because of the plasticity of food practices.

However, there are examples where food practices are not quite diversified and combine almost systematically the same ingredients. This is the case, for example, of the "Thiebboudiene"[9] in urban areas in Senegal, which most often associates small pelagics (round or flat sardinellas) and rice, which can then be considered as complements.

The elasticity of demand ε_d is also a very important concept for the economic analysis of markets. It is a measure of the sensitivity of the demanded quantities to price variations. It is mathematically defined by equation [3.10]. The flexibility of prices, which is the sensitivity of prices to the variations of the quantities, is the inverse of the elasticity of the demand:

$$\varepsilon_d = \frac{\frac{\Delta Q}{Q}}{\frac{\Delta P}{P}} = \frac{\Delta Q}{\Delta P} \cdot \frac{P}{Q} \qquad [3.10]$$

In the case of a linear demand function, the elasticity is variable along the demand curve (Figure 3.6); it is located between $-\infty$ and 0. In the case of a demand function of the type $P = A.\ Q^{-\beta}$, the elasticity will be constant and equal to $-\beta$ and the flexibility to $-(1/\beta)$. The case of unit flexibility in prices is particularly interesting because any reduction in the quantity of fish is then compensated by a proportional increase in price, therefore enabling the incomes of producers to be maintained as constant, for example in the case of a quota policy implementation.

Different types of market situation can be observed, which will affect the shape of the demand function for fishery products:

– the isolated fishery with a narrow market: this situation is observed in countries where internal trade and exports are poorly developed due to the lack of marketing infrastructure for fishery products or even the limited interest of consumers for fish. The slope of the demand function will then be significantly negative. A

9 Wolof (main language used in Senegal) expression to designate the national dish "rice with fish", which is very popular and the most current in daily meals.

borderline case, but not an unrealistic one, is the one in which the market cannot absorb the maximum equilibrated production (MSY). This situation was formerly that of the fisheries of small coastal pelagics from West African countries, such as Mauritania, Morocco and Senegal whose absorption capacity of the local market (Mauritanian or Moroccan cases[10]) was limited by the demand of consumers or by poor transport infrastructures (case of Senegal until the 1960s) and did not allow the distribution of the potential national production. This situation now tends to disappear with the development of the exportations from the countries whose domestic demand is still insufficient or through the development of internal marketing infrastructures for those whose domestic demand was significant but unsatisfied (Senegalese case);

– the isolated fishery with a broad market: in this configuration, the landed quantities will always influence the prices, but the market can absorb the maximum quantities that can be landed. This has long been the case of demersal species in West Africa before the development of exports to Europe or Asia;

10 The Mauritanian and Moroccan cases are very interesting. These two countries with abundant fishery resources, notably pelagics (sardines, sardinellas, horse mackerel, etc.), have long been characterized by a weak national demand for sea products, due to consumer preferences in favor of proteins mainly supplied by the farming of ovine and caprine animals. This situation has encouraged the development of the canning industry in Morocco (the ports of Safi and Agadir) intended primarily for export. In Mauritania, the large-scale exploitation of pelagic fishes is effected in the framework of fisheries agreements or with the countries of Eastern and Western Europe (which is equivalent to exporting fish by delegating the direct exploitation to the importing country) and then through *joint-ventures* with China (more recently). This situation could change with the evolution of local markets due to changes in consumer preferences (hence the development of specialized catering in sea products can be seen in Morocco, the population of Southern Mauritania consumes more fish of marine origin), the relative prices of meat and fish (the price of meat has a tendency to grow faster than that of fish because of the difficulties in traditional breeding), and the emergence of a regional market (the Senegalese market absorbs, for example, more and more products coming from Mauritania, which can be caught by Senegalese fishermen following the bilateral agreements or even exported through the intermediary of wholesalers). Finally, landlocked countries such as Mali and Burkina Faso are increasingly importing sea products in their transformed form (dried and smoked) and even fresh or frozen.

– price-taker fishery (infinite elasticity of the demand): this is the case of export fisheries that sell, on the international market, a quantity that can be regarded as marginal at the scale of the worldwide demand. The supply of the individual fishery cannot influence the world price, which is imposed on it. Multiple examples exist, such as tropical shrimp fisheries (Madagascar, Gabon, Senegal, Guyana, etc.) or even tuna fisheries to a lesser extent. In this case, the demand function for a particular fishery will have a zero slope and the demand will be infinitely elastic. This case is often retained in the most simple presentation of Gordon–Schaefer's model (model known as "with a constant price");

– open export fishery with influence on the formation of prices: this is a relatively rare, but very interesting, case of fisheries with strong exportations and whose production is sufficiently important to influence the world price. In this case, the demand function for fish again shows a negative slope. Two examples can be presented: the fishing of octopus (*Octopus vulgaris*) in Morocco with 15% of the world supply of octopus, including all species[11], and the Humboldt current anchovy fishing (Peru and Chile) with 50% of the world supply intended for the production of fish meal [MER 10].

3.2.1.3.2. The cost function

In the standard bioeconomic model, the total operating cost CT is a linear function of the fishing effort E, c being the unit cost of the effort:

$$CT = c.E \tag{3.11}$$

It is considered that the unit cost of the effort takes into account the opportunity cost of capital and labor, that is to say the revenues that could be expected from them if they were used in an alternative activity[12]. In addition, the amortization of the capital (renewal due to physical wear and obsolescence) is taken into account.

11 According to FAO FishStatJ (www.fao.org/fishery/statistics/software/fishstat/fr).

12 With regard to the capital, the interest rate on a low-risk bank investment is often retained.

This formulation, however, generates a few comments and some criticism. It is considered that the full cost is variable, which seems logical for a long-term model which considers all the factors of production as variable. This argument may, however, be discussed because fisheries often require a fixed shore-based infrastructure (landing wharfs, auctions, storage facilities, etc.) totally independent of the fishing effort, even in the long run. The fishing effort, which is the only attribute of the cost function, constitutes the second critical point. Reality often appears to be different and more complex. Some cost elements are linked to the duration of fishing trips: fuel, and notably crew food. The cost of labor is shaped differently. Several modes of labor remuneration are possible: proportional to the volume or to the value of the catch (eventually with a fixed element), or even to the value of the catch deducted from the common charges (fishing trip costs) shared between the crew and shipowner. Costs of landing and fish freezing are finally proportional to the catch. A better adapted form of the cost function would, therefore, be:

$$CT = CF + c_1.E + C_2\left(P\left(Q(E)\right).Q(E), E\right) + C_3\left(Q(E)\right) \qquad [3.12]$$

where CF represents the fixed costs, c_1 is the unit cost of charges proportional to the fishing effort, C_2 is the function cost of labor and C_3 is the cost function of landing and freezing.

In the firm theory, the average cost CMo is equal to the total cost CT divided by the catches Q, and it must be distinguished from the unit cost of the effort c (equation [6.12]). The marginal cost CMa is defined as the variation of the total cost CT associated with the production of an additional unit of fish. From a mathematical point of view, this is the first derivative of the function of total cost with respect to the catch:

$$CMa = \frac{\partial CT}{\partial Q} \qquad [3.13]$$

If the average and marginal costs (denoted as $CMoe$ and $CMae$) are no longer expressed as a function of catches Q, but of effort E, it can be deduced from equation [3.11] that they are equal to c. The two formulations of average and marginal costs are used in the different possible presentations of Gordon–Schaefer's model.

3.2.1.4. *The economic results of fish exploitation*

In the standard version of the model, it is considered that the creation of wealth by the fishery, called economic rent[13] and denoted by R, is equal to the profit Π, which is the difference between the income (product of the catch by the price) and the cost:

$$\pi = Q(E,X).P(Q(E,X)) - c(E) \quad [3.14]$$

It can be assumed that the model is here also very simplifying because it leaves aside the question of sharing fishery economic rent between private fishery operators (who perceive a benefit) and the rest of the economy that can levy taxes T on the economic results of the fishery or even, on the contrary, pay subsidies S to support the sector. The only control variable determining the profit here is the fishing effort; it is considered that the fish market is outside of the capacity of action of the fishery manager. If we take into account taxes and subsidies, economic rent is then different from the profit:

$$R = \pi + T - S \quad [3.15]$$

Finally, in order for the bioeconomic approach of fisheries to agree with the conventional macroeconomic approach for the measurement of the economic wealth, the concept of net added value (*NAV*)[14] should be used and no longer the rent. The *NVA* is equal to the sum of the profit Π, labor income W and net transfers *(T-S)* with the rest of

13 The article "rent" in the *New Palgrave Dictionary of Economics* gives the following definition: "'Rent' is the payment for use of a resource, whether it be land, labor, equipment, or even money. Typically, the rent for work is called 'wages'; the payment for land or equipment is often called 'rent'; and the payment for use of an idea is called a 'royalty' and the payment for use of money is called 'interest' [...]. To early economists, 'rent' meant payment for use of land" [ALC 87].

14 It will be recalled that the gross domestic product (GDP) is equal to the sum of the Net Added Value of all the sectors in the national economy, net of taxes.

the economy. Unlike the previous approach, it no longer considers the remuneration of the work W as a cost but as a creation of wealth[15]:

$$NAV = \pi + W + T - S \qquad [3.16]$$

This point is very important especially when fisheries are characterized by a low capital per capita (labor using technologies), such as in the majority of tropical artisanal fisheries. Some authors consider that the objective of such fisheries is mainly the contribution to employment and wages and then to durable livelihoods for fishing communities [BEN 10] rather than the only creation of economic rent.

3.2.1.5. *Fishery behavior in free access*

Here, the standard model is being used in its simplest version, that is with a linear cost function identical to that of equation [3.11] and a demand function infinitely elastic (constant price independent of the landings). The economic rent R is here equal to the profit Π. The model is presented in Figure 3.7. The income curve is a simple vertical translation of the catches curve in Figure 3.4. It passes through a maximum that corresponds to the *MSY*, expressed here in value and no longer in volume, for the same effort level E_{msy}. The rent (or profit in this case) is equal to the vertical difference between the curve of income and the straight line of the costs *CT*. It passes through a maximum *MEY* which is the maximum economic yield – MEY), for a effort level E_{mey}.

In the absence of barriers to entry[16] in the fishery, "the tragedy of the commons" previously discussed (see section 3.1.2.2) is observed and the myopic behavior of fishermen drives the fishing effort to develop, as long as a positive individual profit is possible. This leads

15 The choices referred to are not only scientific: considering the remuneration of the capital as a positive outcome and that of labor as a cost is based on a conception of the economic world which not everyone agrees with ... even if it tends to become hegemonic within the economic discipline.

16 In economics, the phrase "barriers to entry" means all the constraints that limit the free movements of men and capital at the entry or exit of a sector. The standard model of pure and perfect competition considers that the absence of barriers to entry is one of the conditions for the optimal allocation of resources within an economy. The fisheries sector is an exception, among others, to this proposal.

them to invest until a free access equilibrium (also called bioeconomic equilibrium or bionomics in the literature) corresponds to point A in Figure 3.7. This equilibrium is stable (on the right, it is in a situation of economic loss and the effort tends naturally to decrease; on the left, the profit is positive and the effort increases). It is obtained for an effort level equal to E_∞.

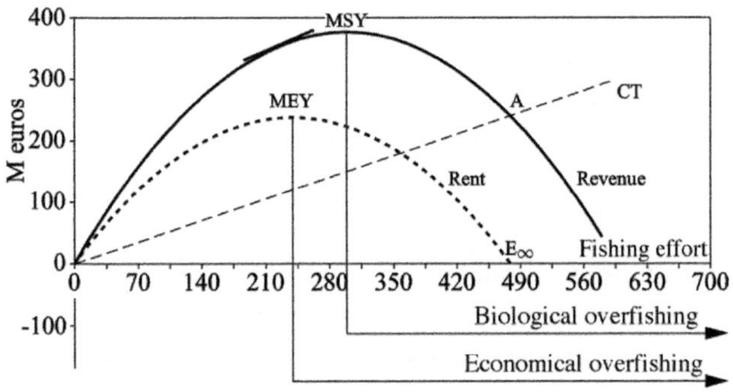

Figure 3.7. *Gordon–Schaefer's bioeconomic model*

It should be noted that the position of economic optimum is also determined by the equality between marginal cost and revenue of effort, which is a usual condition for obtaining an optimal production in economic theory. The rent is maximized when its first derivative with respect to fishing effort is equal to zero; therefore[17]: $\frac{\partial R}{\partial E} = 0$.

As $R = REV\text{-}CT$, where REV is the income, this condition is equivalent to $\frac{\partial REV}{\partial E} = \frac{\partial CT}{\partial E}$ and therefore to the equality between marginal revenue and the cost of effort.

Two definitions of overfishing are deduced from this model. Economic overfishing is a situation where the fishing effort is higher than that which makes it possible to reach MEY. Biological

17 To be rigorous, it should also be necessary to verify that the second derivative is negative, but the fulfilment of this condition is obvious because of the concavity of the rent curve.

overfishing is reached when the fishing effort is higher than that which makes it possible to achieve MSY. In both cases, overfishing corresponds to a waste of economic resources (human and technical resources are allocated in large numbers in the fishery sector while they could be used more efficiently in other sectors) as well as to a reduction in the abundance of the exploited stock, which from an economic point of view is also a waste of natural capital (but, in this model, without risk of extinction).

3.2.1.6. *Comments and discussions on the standard Gordon–Schaefer's model*

The economic optimum defined here is rather more conservationist than the biological optimum because it leads to advocation of a lower level of fishing effort to which corresponds a greater abundance of exploited biomass. This contradicts the assertions too often made about the responsibility of economic recommendations with regard to overfishing.

An avid reader of economics may however be surprised that the cancellation of the long-run profit, generally regarded as a factor of efficiency in most economic sectors[18], does not also correspond to an optimum in the case of fisheries ... This can be explained because the main factor of production, the biological resource, is not owned here (absence of well-defined property rights) and therefore unpaid. The comparison can be made with agriculture, where production relies largely on a natural resource (the fertility of the land), but which differs completely from the fishery according to the status of the natural capital, which is owned here, remunerated (depending on the land tenure system) and maintained (reconstitution of fertility either by fallowing or addition of artificial nutrients). This is far away from the situation of the free access fishery where man maintains a simple extractive relationship with the resource, whose nature has not radically changed, compared to that prevailing at the time of the prehistoric hunter-gatherers, but today with different and more efficient technical means.

18 Although the search for the maximum individual profit explains the behavior of producers, competition within the sector (and in particular the absence of barriers to entry) leads inexorably profit to be dissipated within the sector by the arrival of new firms "attracted" by the possibility of economic gains.

3.2.1.7. *Responses to price changes*

Concerning the responses of the model to price changes, it is necessary to distinguish between those relating to a price change with an infinitely elastic demand (price taker fishery) and those linked due to a shift of a demand curve with negative slope.

Figure 3.8 shows that a price increase (in this example, from 50 to 75 euros per kg) induces in a mechanical way a vertical translation of the income curves (from Rev1 to Rev2) and rent (rent1 to rent2). The increase in the price, from the initial bioeconomic equilibrium A, allows a temporary improvement of profit and, therefore, an increase in fishing effort, leading the fishery toward a new free access equilibrium B. In addition to the increase in fishing effort, the price increase will also result in a biomass decrease. In the absence of fishing effort control, any policy designed to increase the ex-vessel price will only produce positive short-run effects and will lead to undesirable effects in the long run.

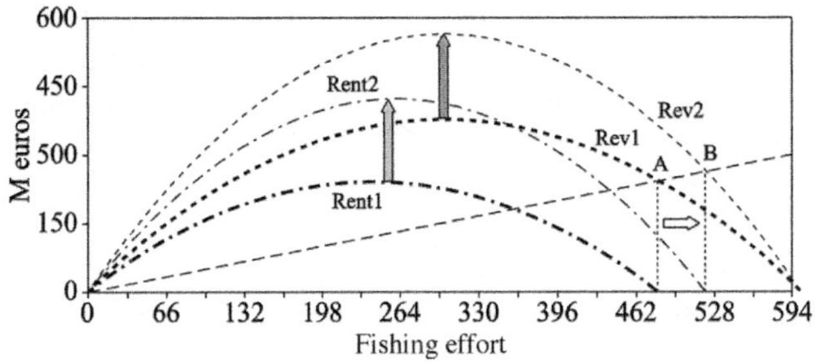

Figure 3.8. *Response of the Gordon–Schaefer's model to price increase*

We will now study the consequences of the shape of the demand function on the bioeconomic equilibrium and the optimal reference points. From the different market contexts discussed above, we will retain four cases (Figure 3.9):

1) horizontal demand function (model with a constant price, price taker fishery);

2) linear demand function with a negative slope with *P(MSY)* > 0, with an ordinate at the origin equal to that of the first case;

3) demand function with a negative slope with *P(MSY)* < 0, with an ordinate at origin equal to that of the first case;

4) demand function with a negative slope with *P(MSY)* < 0, with an ordinate at origin three times that of the first case.

Figure 3.10 shows that the bioeconomic equilibrium and the reference points (MEY and MSY in value) are significantly affected depending on the shape of the demand function retained. Table 3.1 indicates the results obtained according to the cases. The comparison of cases 1 and 2 shows that the change from a function with a zero slope to a function with a slightly negative slope first modifies the shape of the income curve. In the first case, the MSY expressed in value is a multiple of the one expressed in volume, for the same fishing effort. In the second, the MSY value is lower and obtained for two different values of the fishing effort. The MEY is also lower and obtained for a reduced effort of more than 50%. The bioeconomic equilibrium is reached for an equally much lower fishing effort. These differences are more important in case 3 where there is also observed a range of fishing effort for which the income is zero. The last case (4) leads to even more confusing results: three bioeconomic equilibriums are possible (A, B, C), two of which are stable (A and C) and one unstable (B). Imagine that the fishery is impacted by some change in the demand function that leads to case 4. The final bioeconomic equilibrium will then depend on the initial situation: this will be point A if we start from case 3 or point C if we start from case 1.

Case	Demand function parameters		MSY in value	E_{MSY}	Rent MEY	E_{mey}	Effort at bioeconomic equilibrium
	a	b					
1	50	0	375	295	239	250	480
2	50	–4	156	165 419	89	99	300
3	50	–7	88	75 505	54	55	135
4	150	–25	400	99 505	350	74	250 380 560

Table 3.1. *Different cases of demand functions and their consequences on the bioeconomic equilibrium and reference points of the standard Gordon–Schaefer's model*

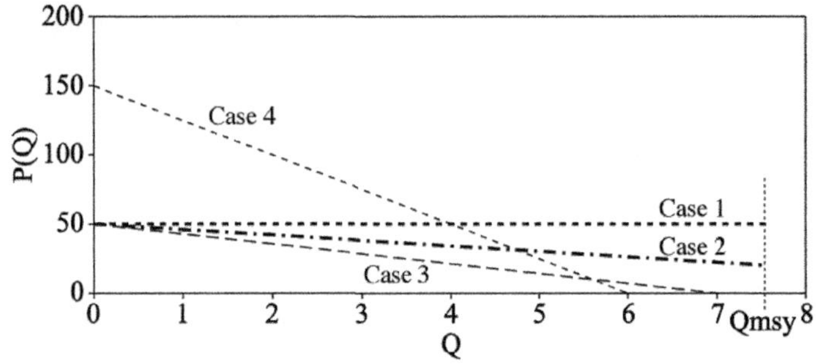

Figure 3.9. *Different cases of demand functions*

These various examples show that the demand conditions have very important consequences on the economic results of the fisheries and their equilibriums. They emphasize the necessity of obtaining correct information on the market, which is unfortunately not always the case. Finally, market changes can also be as destabilizing as those from the biological components of the fisheries.

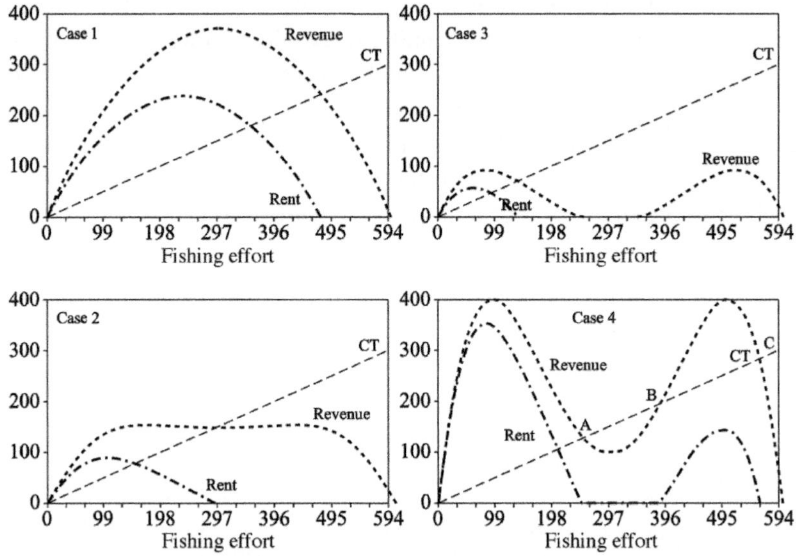

Figure 3.10. *Impact of the demand function on the bioeconomic equilibrium in the standard Gordon–Schaefer's model*

3.2.1.8. *Responses to cost changes: consequences of subsidies or taxes*

Any modification of the unit cost *c* causes a change in the slope of the total cost function *CT* and leads to changes in the position of the bioeconomic equilibrium as well as the economic optimum MEY. Reducing by 20% the unit cost *c* with respect to the reference simulation leads to an increase from 480 to 505 in the effort at bioeconomic equilibrium (Figure 3.11).

If the fishery is managed to remain at the economic optimum, this cost reduction would lead to an increase in the MEY from 240 to 265 M€, for a quite equivalent fishing effort. We can, therefore, conclude that a decrease in cost will not improve economic results if fishing effort is not regulated, whereas it allows a profit increase if effort is maintained below bioeconomic equilibrium.

The situation becomes more complicated when we want to understand the consequences of subsidy or taxation measures. The latter may apply to the fishing effort in which case they will affect the slope of the cost function. On the other hand, if they are applied to catch or income, they will change the income curve.

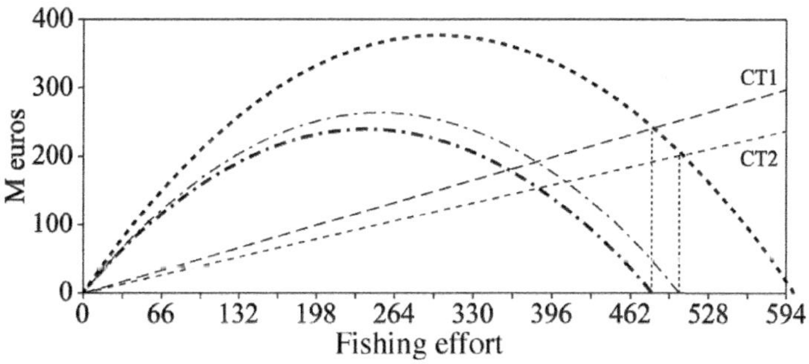

Figure 3.11. *Impact of a unit cost reduction of the effort on the bioeconomic equilibrium in the standard Gordon–Schaefer's model*

A taxation of the fishing effort will lead to an increase in the fishing cost and therefore to a new bioeconomic equilibrium due to reducing the equilibrium effort. On the other hand, unlike the previous example, this new free access equilibrium no longer leads to the total dissipation of the economic rent because the taxes levied by the state constitute an element of economic rent as previously stated (equation [3.15]). Conversely, a subsidy of fishing effort (Figure 3.12) leads to a temporary situation of profitability (the total cost shifting from CT1 to CT2) and encourages an increase in fishing effort up to a new equilibrium B. In the absence of control of fishing effort, the subsidy therefore leads to a negative rent in B: the fishermen do not obtain more profit than in A but the rest of the society has assigned additional public resources to fisheries (negative income of the state). If the fishery is managed at the economic optimum, the subsidy, on the other hand, has a neutral effect on the total wealth: the increase in profit is just compensated by the financial contribution of the state, but it also leads to a redistribution of wealth in favor of the private sector.

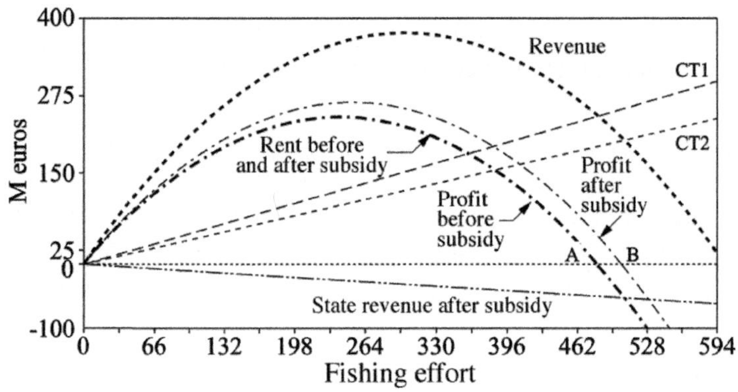

Figure 3.12. *Impact of a fishing effort subsidy on the bioeconomic equilibrium in the standard Gordon–Schaefer's model*

Catch taxation (Figure 3.13) affects the income curve as well as the profit curve, and leaves unchanged the cost function; the consequence will be a reduction in the bioeconomic equilibrium effort from A to B and a downward shift of the income curve.

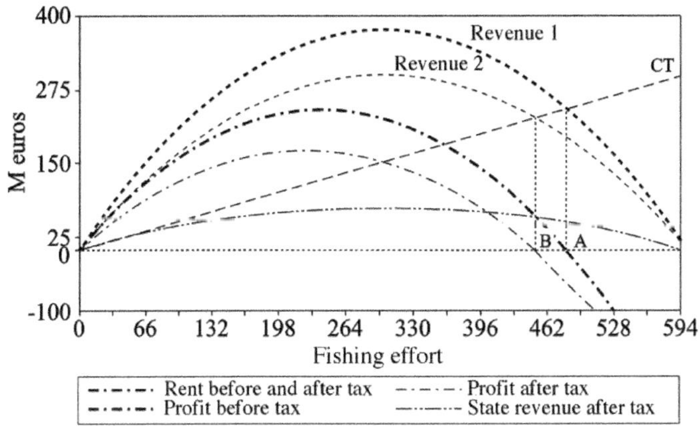

Figure 3.13. *Impact of catch taxation on the bioeconomic equilibrium in the standard Gordon–Schaefer's model*

The rent curve remains unchanged, and there appears to be a positive rent at the new bioeconomic equilibrium B, which is equal to the tax levied by the state.

3.2.1.9. *Changes in catchability*

To conclude with the responses of Gordon–Schaefer's model to changes in the parameters, we will deal with the question of catchability, but will not discuss the effect of changes in strictly biological parameters (r and K) which relate to other contributions to this book. As already mentioned, catchability is the result of the behaviors of fish and fishermen. It is particularly affected by the technical choices relating to fishing equipment (mesh sizes, dimensions, etc.) as well as by the increase of its effectiveness[19].

Figure 3.14 presents the consequences of an annual increase in catchability of 2% over a 10 year period.

19 Fishery scientists speak of an increasing trend of drifting of fishing power, which is estimated for some fisheries to be 2 or 3% per year. To this, timely changes must be added, introduced by technical innovations, such as bird radars, fish-aggregating devices in tropical tuna fishing, or the use of GPS in artisanal fisheries. The phenomena of learning by professional experience, although difficult to assess, also plays an important role.

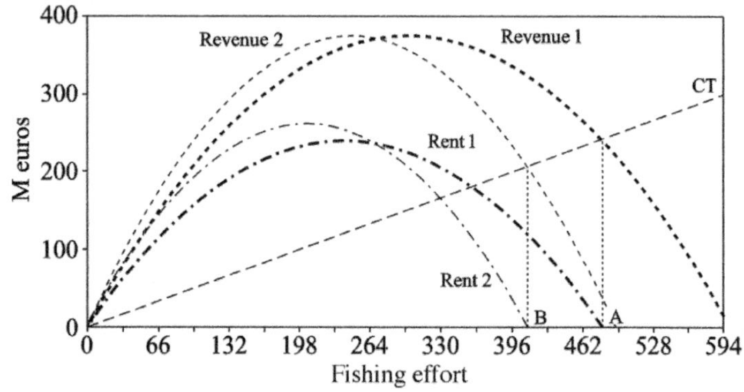

Figure 3.14. *Impact of an increase in catchability on the equilibriums of Gordon–Schaefer's model*

The increase in technical efficiency results in a shift of the curve of income toward the left and the bioeconomic equilibrium moves from point A (fishing effort equal to 480) to point B (400). If the fishery is regulated and maintained at a bioeconomic optimum, this optimum reaches a higher level at the end of the 10 year period with a lesser fishing effort. Contrary to what is often claimed, the increase in efficiency, or technical progress, can contribute to improving economic results if fishing effort is controlled[20].

Conversely, the measures destined to reduce the technical efficiency (inverse reasoning of the previous), such as those aimed at modifying the fishing equipment to reduce fishing mortality, do not result in economic gains. This result is, however, strongly dependent on the hypotheses of the model. Other model formulations taking into account the effects of selectivity on the size of fish caught can give other results, as will be seen later.

20 Many authors, notably fisheries biologists, have a systematically negative vision of technological progress due to the increase in fishing power that results from it. This leads us to encourage the reduction of the effectiveness of the fishing units (e.g. by limiting the size of equipment for each fishing unit). From an economic point of view, this reasoning is questionable: it is better to have a limited number of very effective fishing units rather than a higher number of units with low fishing power. Some other considerations, notably social, can result in criticizing this economic point of view, which we favor here.

3.2.1.10. *Extension of the model to the case of the joint exploitation of several stocks*

One of the main conclusions of the standard model is that the uncontrolled exploitation of fish resources leads to very poor economic results and reduces the level of biomass well below that of the optimum exploitation. However, there is no risk of total resource exhaustion because there is equality between catch and natural growth for any value of fishing effort. It will be seen later that if this assumption of equilibrium is abandoned, the model behavior will radically change. However, even maintaining the hypothesis of equilibrium exploitation, it is possible to demonstrate the possibility of resource exhaustion due to fishing, when several stocks, with very contrasted economic and biological characteristics, are exploited simultaneously.

In the following example, the joint exploitation of two very different stocks by the same fishing technique is presented (Table 3.2). It shows the trawl exploitation of the coastal penaeid shrimp which also impacts a by-catch fish resource [CHA 11a]. While shrimps are first characterized by very high intrinsic growth (r), fish have a slower growth rate r but a much larger carrying capacity K.

	r	K	q	P
Shrimps	0.5	10,000	0.0004	50
Fish	0.2	50,000	0.0006	10

Table 3.2. *Joint exploitation of shrimps and fish; parameter values of the simulation*

Catchability q is higher for fish because of its greater size; it is more vulnerable than shrimps to equipment designed to catch priority in prawns. Finally, shrimps, mainly sold on the world market, have a much higher price than fish. The cost parameter c is obviously identical for the two species that are caught with the same technology. Figure 3.15 shows the response of the two species to the joint exploitation in terms of caught quantity (Q) (top), biomass (bottom) and economic outcomes (intermediate graph). The bottom figure shows the equilibrium linear relationships between fishing effort and biomass (see equation [3.8]) for the two species. The catches and income of the fishery are the sum of those obtained for each species, hence their particular form with two local maximums. If the fishery is unregulated (free access), the bioeconomic equilibrium is point A in the intermediary figure for a fishing effort equal to 625. The very important point here is that this equilibrium is obtained for a biomass with a zero value for species 2 (fish) which is therefore exhausted, whereas that of species 1 (shrimp) still represents 20% of its initial value. If the fishery is exploited at the economic optimum (point B), species 1 is still driven to near extinction. One of the important results (which lends itself to criticism) of the standard monospecific model, in other words the inability to exploit a stock until its disappearance, is therefore called into question when the exploitation of several species is taken into account.

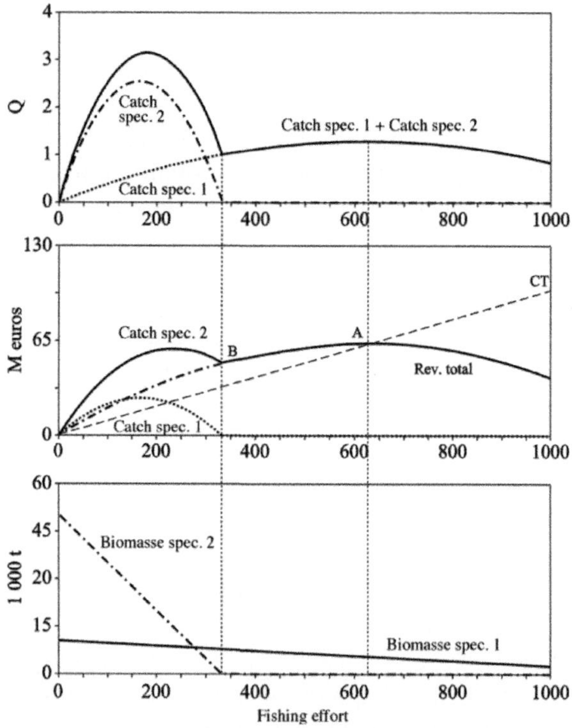

Figure 3.15. *Joint exploitation of two species (shrimp and fish) with very different biological and economic parameters*

3.2.2. *Dynamic non-equilibrium model*

The main hypothesis of the previous model is equilibrium between the natural growth $G(X)$ and the catch $Q = q.E.X$ for any value of fishing effort. It is also what economists call a comparative static[21] model where the characteristics of the different possible equilibriums are discussed without questioning the system trajectory from one equilibrium to the other.

We will now present a discrete model, inspired by Smith's continuous model [SMI 69] which is no longer based on the equilibrium hypothesis between catches and growth. In addition, this model represents more explicitly the relationship between profit and fishing effort by introducing an investment function.

3.2.2.1. *The presentation of the model*

The growth of the biomass is written as in the previous model, but the state variable describing biomass X is henceforth also a function of time t:

$$G\left(X(t)\right) = r.X(t).\left(1 - \frac{X(t)}{K}\right) \qquad [3.17]$$

The catch Q is calculated in a similar manner as in the previous model:

$$Q(t) = q.E(t).X(t) \qquad [3.18]$$

Biomass X at time t is equal to the sum of the biomass and its growth minus the catches in t -1:

$$X(t) = X(t-1) + G\left(X(t-1)\right) - Q(t-1) \qquad [3.19]$$

---

21 In economics, comparative static is an "approach which consists of comparing the equilibriums of a model before and after one of its parameters has been changed, this comparison is static because it says nothing of the way in which the changeover from one equilibrium to another is effected" [GUE 02, p. 994].

Total cost CT, profit π and rent R are obtained in an identical way to the standard model:

$$CT(t) = c.E(t) \qquad [3.20]$$

$$P(t) = a - b.Q(t) \qquad [3.21]$$

$$\pi(t) = P(t).Q(t) - CT(t) \qquad [3.22]$$

$$R(t) = \pi(t) + T(t) - S(t) \qquad [3.23]$$

The second peculiarity of this model is the investment function. At each time period, the net variation of economic capital in fishery, that is to say the investment I, is proportional to profit. If profit is positive, its share going to investment will be equal to s_1; if it is negative, it will be equal to s_2. This difference is introduced because it is considered that the behavior of entry into the fishery may not be symmetrical to that of exit, considering that fishermen invest rapidly during periods of prosperity and are rather more reserved to come out when things are deteriorating, each hoping that others will leave before him and that he will thus be able to benefit from the resulting improvement[22]:

$$I(t) = s_1.\pi(t) \text{ if } \pi(t) > 0 \text{ or } I(t) = s_2.\pi(t) \text{ if } \pi(t) \leq 0 \qquad [3.24]$$

The fishing effort is determined by the following equation:

$$E(t) = E(t-1) + I(t-1) \qquad [3.25]$$

3.2.2.2. *Behavior of the dynamic model in discrete time and comparison with the standard equilibrium model*

Figure 3.16 presents the equilibriums and trajectories of the discrete dynamic model and compares them with that of Gordon–Schaefer's model (in the left part of the figure). The right part of the figure shows the evolution of the biomass and the fishing effort, with the dynamic

22 This is another application of the prisoner's dilemma by Nash in the case of fisheries.

model. The common parameters of both models are presented in Table 3.3; only s_1 and s_2 are specific to the dynamic model. The two models use a constant price (horizontal demand curves).

Parameters	r	K	Q	a	b	c	s_1	s_2
Value	0.3	100	0.0005	60	0	1	0.1	0.05

Table 3.3. *Parameters used for the comparison of the dynamic model in discrete time and the Gordon–Schaefer's standard model*

The bioeconomic equilibrium A is obviously identical for the two models, but the trajectories followed to achieve it are quite different. During the developing phase of the fishery (increasing effort), the trajectories of the income (those of catches are similar) and rent are higher than that of the equilibrium model. The kinetics of the dynamic system leads the fishing effort far beyond its free access value in A (400) and the biomass below its equilibrium value (33). It is also important to emphasize that the maximal rent observed during the development phase greatly exceeds the maximal equilibrium rent.

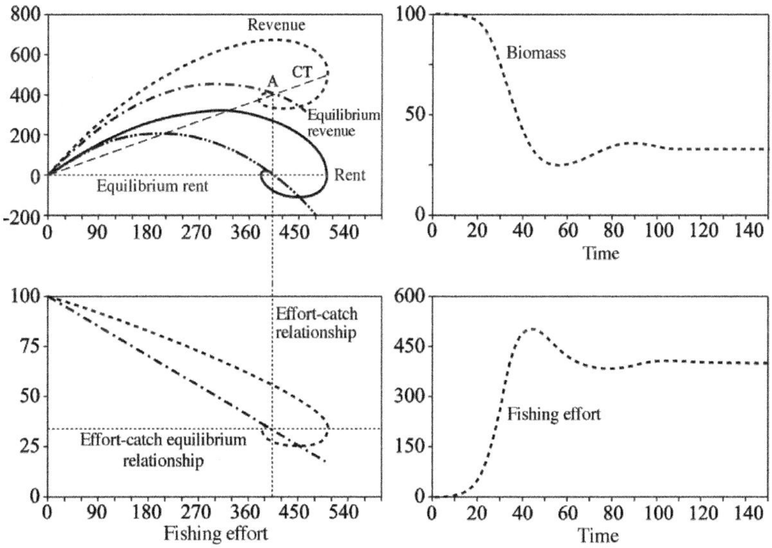

Figure 3.16. *Comparison of the equilibriums and the trajectories of Gordon–Schaefer's standard model and Smith's dynamic model in discrete time*

In this new model, the bioeconomic equilibrium always plays its role of attractor, but the trajectories to reach it no longer respect the equilibrium constraint between growth and catch. The time necessary to achieve an equilibrium will be a function of the parameters of the resource growth r and investments s_1 and s_2. Finally, any changes in the economic or biological parameters will be a source of instability and it is necessary to know whether instability will vanish or not in time.

3.2.2.3. *Dynamic model responses to changes in economic parameters*

Various simulations enable exploring the responses of the dynamic model with respect to a reference simulation. Their parameters are presented in Table 3.4.

Simulation no.	Parameters	r	K	q	a	b	c	s_1	s_2
1	Reference simulation	0.3	100	0.0005	40	0	1	0.1	0.05
2	Change in price	0.3	100	0.0005	**100**	0	1	0.1	0.05
3	Change in the demand function	0.3	100	0.0005	**100**	**-2**	1	0.1	0.05
4	Decrease in the unit cost of the effort	0.3	100	0.0005	40	0	0.5	0.1	0.05
5 5.1	Change in the investment function	0.3	100	0.0005	40	0	1	**0.5** **1**	0.05
6 6.1	Change in price and in the investment function	0.3	100	0.0005	100	0	1	**0.5** **1**	0.05
(The modified parameters in each scenario appear in bold)									

Table 3.4. *Parameters of the simulations of changes in economic parameters in the discrete time dynamic model*

Simulation no. 2 in Figure 3.17 shows the model response of an increase in fixed price (in other words, it is independent of catches) from 40 to 100. The coordinates of the equilibriums are modified in an identical manner to the standard model (the equilibrium effort increases by 60 % compared to the reference situation and the equilibrium biomass is reduced by 60%). In contrast, the trajectories of state variables in simulation no. 2 deviate greatly from the values of final equilibrium to which they converge.

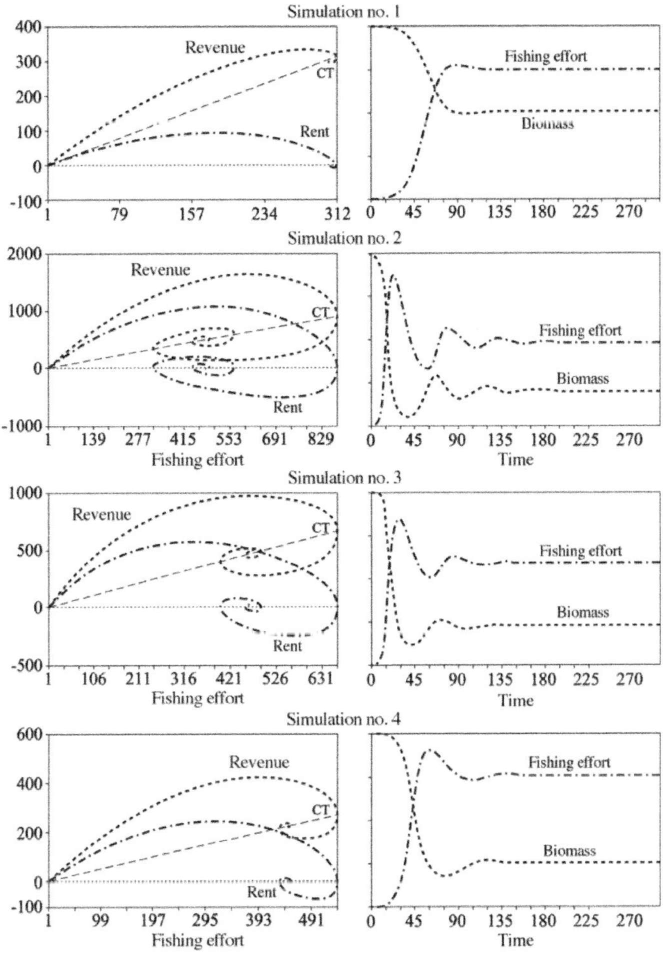

Figure 3.17. *Responses of the dynamic model to changes in parameters*

Thus, the maximum value of the effort (874) is 82% greater than that of equilibrium (300) and the minimum value of the biomass is 25% lower than that of equilibrium (20). These results suggest the role of price variation as a source of fisheries instability.

These results are mitigated when prices are no longer independent of catches (simulation no. 3). In this case, the decrease in price associated with any increase in catches attenuates the magnitude of the oscillations (the opposite effect occurs in the event of a reduction of catches). This result is important because it emphasizes that the case of "price takers", fisheries fostered by the development of the international trade of fishing products, is potentially more unstable than that of fisheries supplying a local market (and whose demand function has a negative slope).

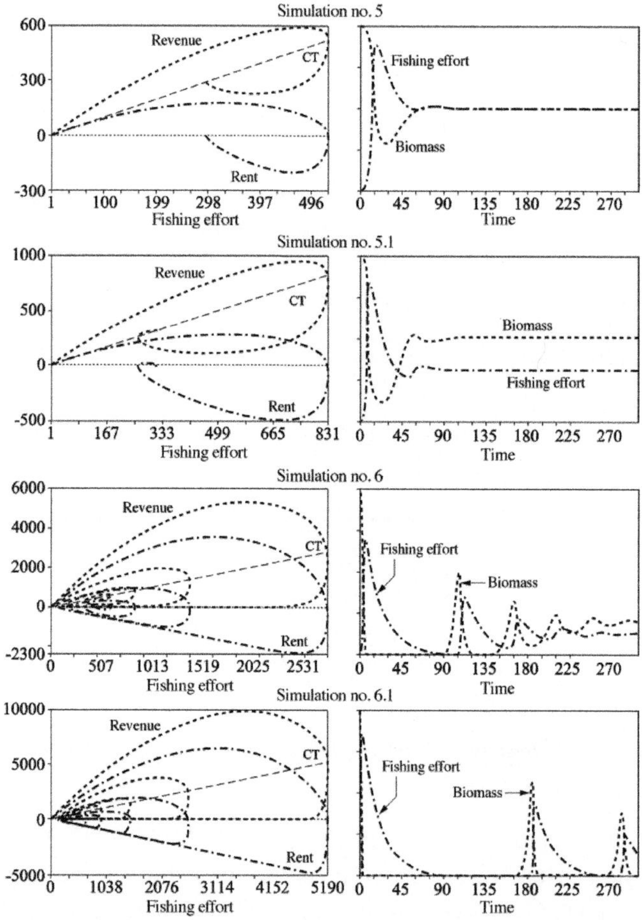

Figure 3.18. *Response of the dynamic model to changes in parameters (continued)*

The reduction of 50% in the unit cost of fishing effort (simulation no. 4) leads to a result close to that of the increase in the price. It leads to an increase of 50% in equilibrium effort and to a similar reduction in the equilibrium biomass, and there is convergence toward these values after some oscillations. Any reduction in cost or increase in subsidy leads to a decrease in biomass and an increase in equilibrium effort and reinforces the instability of the fishery.

The following two simulations (5 and 5.1) show the consequences of changes in the investment function: the share of reinvested profit (s_1), initially equal to 0.1, shifts successively to 0.5 and 1 (Figure 3.18), whereas s_2 is not modified. This does not lead to modifications in the equilibrium values of effort and biomass. On the other hand, their values pass through extrema which deviate substantially from the values of equilibrium before converging to them (+76 and +177% for the fishing effort, −4 and −78% for biomass). Any change in the investment function results in modifying the magnitude of the fluctuations of the fishery.

The last two simulations (6 and 6.1) resume the same hypotheses as the previous for the investment function by adding the double of the fixed price p. This modification strengthens the unstable behavior of the fishery, and the convergence toward the equilibrium is reached after a significant number of periods. Finally, and in particular in the case of simulation 6.1, an alternation is observed of near fishing stoppage followed by periods of very rapid development. The time lag between the dynamics of effort and biomass is explained by the dissymmetry in behaviors of entry (very fast) and exit (slower) in the fishery.

3.2.3. *The Clark Munro's model: the golden rule of the intertemporal allocation of natural capital*

3.2.3.1. Natural capital and discount rate

This very economic, even financial, approach considers the fish stock as an asset whose management is a strategy of intertemporal allocation between immediate and future uses. At first approach, it can be assumed that the manager of the natural asset is exploiting the resource in such a way that its biomass is led to a level X_* such that the marginal financial return of this asset is equal to the interest rate i prevailing in the economy (which is equal to the marginal return of the capital in the economy as a whole). i is also equal to the discount rate[23] which makes possible the calculation of the present value of future income.

The marginal return of the resource is equal to the variation of the catch dQ permitted by investment[24] in an additional unit of resource, which is equal to $G'(X)$ if the exploitation respects the equilibrium condition between catch and stock instantaneous growth. A graphical presentation helps illustrate this reasoning (Figure 3.19).

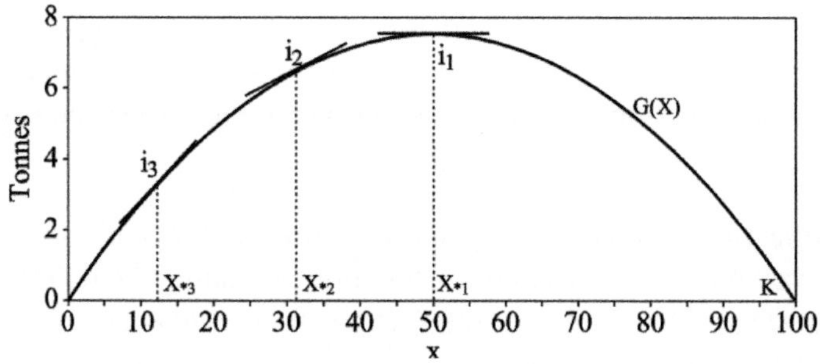

Figure 3.19. *Stock optimal level as a function of the discount rate*

23 The discount rate is a very important concept, unfortunately often misunderstood by non-economists. The discount is based on two fundamental principles. First, the preference for the present: an individual confronted with two alternatives (without risk) which are to have the same amount of money either immediately or in the future (e.g. in a year) will choose the first solution, because waiting has a cost (the price of time). To this pure preference for the present, a risk premium is added when faced with the intrinsic uncertainty of the future. To return to the previous example, there is a risk of disappearance or erosion of the value of the asset, and the individual himself can disappear during the waiting period ... Experimental economics easily allows us to estimate individual discount rates. To achieve this, it suffices to ask an individual what is the minimal sum x_1 that will have to be paid to him/her in a year in order for him/her to agree to lend (which is equivalent to not use) the sum x_0 during the same period, the resulting discount rate will be equal to $(x_1-x_0)/x_0$ per year.

24 We are using here the wording employed by C.W. Clark: investing in the resource consists of increasing the size of the stock; conversely disinvestment occurs when the abundance of the resource is decreased.

The optimal size of the stock X_* is determined by the equality between the interest rate (or discount rate) i and the slope of the tangent line (the first derivative) to the growth function $G(X)$. We represent here three situations corresponding to a zero discount rate i_1 (indifference between present and future), a high rate (i_3) and an intermediate rate (i_2). When the discount rate is zero, the optimal biomass is equal to that which allows obtaining the MSY. On the other hand, as soon as the discount rate is positive, this tends to reduce the optimal biomass. Here, the logic is once again quite financial. Let us imagine an initial optimal state ($i = i_2$, $X = X_{*2}$) and an interest rate increase up to i_3. The situation is no longer optimal since the marginal return of natural capital is from now on lower than the expected yield of the capital in the economy. It is appropriate then to reduce the biomass in order to bring it to X_{*3}.

This financial logic is not without problems. This is exposed in Figure 3.20 where a comparison can be seen between the optimal exploitation of two stocks characterized by very different intrinsic growth rates ($r_1 = 0.03$, $r_2 = 0.1$) and the same carrying capacity ($K1 = K2 = 100$), for an identical discount rate.

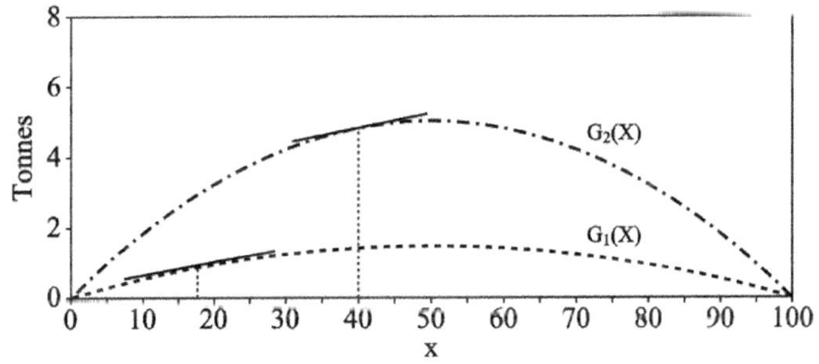

Figure 3.20. *Optimal exploitation of two stocks with very different intrinsic growth rates r*

In the case of the slow growth stock, the optimal exploitation involves a biomass less than 20% of the virgin biomass, whereas it is 40% for the rapid growth stock. An optimal exploitation could then be compatible with the extinction of species with a very slow growth as has been demonstrated in the case of the blue whale [CLA 82].

This model can finally help us to understand the strategies of exploitation of the populations of poor countries subject to economic emergencies and major uncertainties which implies very high implicit discount rates, conflicting with the conservation strategies currently being promoted at the national and international levels [CHA 07][25].

This result questions one of the conclusions of the standard Gordon–Schaefer's model, for which the implicit discount rate is zero, namely the prescription of economists would be systematically more favorable to conservation than biologists (the MEY of the standard model is always obtained for a biomass higher than that corresponding to the MSY). In the model outlined here, the economic optimal exploitation is always less favorable to conservation than that arise from a biological point of view, except in the case of a zero discount rate where they are identical.

3.2.3.2. *The relationship between biomass and cost taken into account, the marginal stock effect*

The analysis of this model may be done in a more formal way, taking into account, in particular, the relationship between the cost of the exploitation per unit of catch c_y[26] and biomass X. We still assume that the price p is independent of catches.

Let us assume that the stock is at its optimal level X_* and that a marginal deviation ΔX occurs. The immediate economic gain obtained from this change will be equal to $-(p-c_y).\Delta X$. A reduction of stock ($\Delta X < 0$) implies an immediate increase in catch and therefore a positive gain, and the inverse in the case of an increase in stock.

25 A factor of the existence of very high discount rates for the poor Southern populations is the near impossibility for them to have access to the financial banking market, which implies borrowing at usurious rates commonly applied by merchants or lenders. Even in Muslim countries where the practice of the usurious interest is socially condemned, it is maintained in a form diverted by a very significant gap between the cash price and future prices in routine transactions. This terrible constraint, which burdens the poor, explains the global success of microfinance which has allowed millions of poor people to climb out of debt for life.

26 c_y is the cost per catch unit that must be distinguished from the unit cost of the effort c, previously used in the exposure of Gordon–Schaefer's standard model.

This immediate economic gain must be compared with the long-term consequences of the change in the abundance of stock X. This long-term effect is due to a change in the value of equilibrium production Q. The equilibrium condition is always respected; therefore, Q is equal to $G(X)$. A biomass change ΔX causes a variation in production equal to $G'(X)\ .\Delta X$, which corresponds to an economic gain of $(p-c_y).G'(X).\Delta X$ over a time period. The long-run economic effect is equal to the previous expression discounted (using the discount rate i) over an infinite time period:

$$\sum_{t=0}^{t=\infty}\left[\left(p-c_y\right).G'(X).\Delta X\right]/(1+i)^t \quad [3.26]$$

This geometric series converges to:

$$\frac{\left[\left(p-c_y\right).G'(X).\Delta X\right]}{i} \quad [3.27]$$

At optimal stock X_*, the two effects of short and long terms must compensate each other, and we obtain the same result that has been previously graphically demonstrated (Figures 3.19 and 3.20), namely the equality between the first derivative of the growth function $G'(X)$ and the discount rate i:

$$-\left[p-c_y\right]\Delta X+\frac{\left[\left(p-c_y\right).G'(X_*).\Delta X\right]}{i}=0 \Rightarrow G'(X_*)=i \quad [3.28]$$

This result is obtained by considering that c_y (the unit cost of the catch) is independent of X. This hypothesis needs to be reconsidered when the catch per unit of effort is proportional to the biomass, which implies that the unit cost of catches is a decreasing function of X or $c'_y(X) < 0$.

The expression of gain allowed by a variation in the biomass ΔX must now take into account the induced change in cost. It is equal to the product of $c'_y(X)$, the stock variation ΔX and the equilibrium production equal to $G(X)$. Equation [3.28] is then modified and becomes:

$$-\left[p-c_y(X_*)\right]\Delta X+\frac{\left[\left(p-c_y(X_*)\right).G'(X_*)-c'_y(X_*)G(X_*)\right]\Delta X}{i}=0 \qquad [3.29]$$

After an easy transformation of the previous equation, the "golden rule" of the optimal management of a fish stock is obtained [CLA 75]:

$$G'(X_*)-\frac{\left[c'_y(X_*).G(X_*)\right]}{p-c_y(X_*)}=i \qquad [3.30]$$

The left-hand side of equation [3.30] may be designated as the intrinsic discount rate of the fish stock, which must be equal to the social discount rate i. The intrinsic interest rate of the stock has two components. The first $G'(X)$ is the instantaneous marginal production[27] of the resource; it can be negative, positive or zero. The second is called the marginal stock effect, according to Clark and Munro. The marginal stock effect is always positive because $c'_y(X) < 0$. This effect implies a higher optimum stock and is therefore more favorable to the conservation of the resource, and this will be more important when the unit cost c is high. In contrast, a higher price p will result in a less optimal level of biomass. The more significant the difference between price and unit cost, the greater the reduction of the optimal biomass will be.

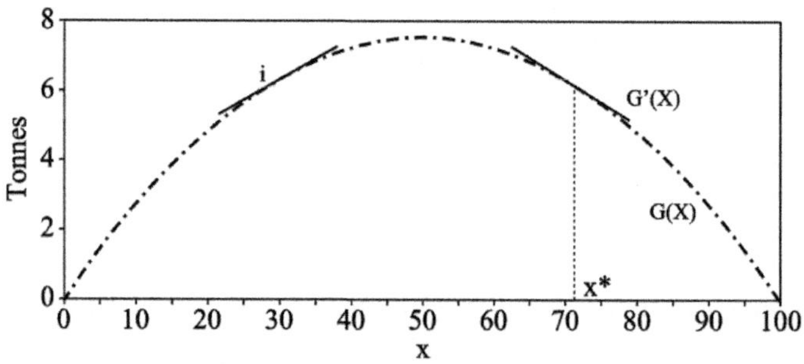

Figure 3.21. *Impact of the stock marginal effect on the determination of the optimal biomass X_**

27 In other words, the additional equilibrium production permitted by a marginal increase ΔX of the fish stock.

The impact of the marginal stock effect on optimal biomass is presented in Figure 3.21. The difference between the interest rate i and the slope of the tangent to the growth function $G(X)$ in X_* is equal to the marginal stock effect.

The case of the two extreme values of the discount rate i must be discussed. In the case of an absolute preference for the present ($i= +\infty$), the optimum is reached for $p=c_y(X)$, therefore with a zero economic rent per unit and a dissipation of the total economic rent. In this case, the optimal biomass X_* is equal to the biomass of free access X_∞, as previously defined.

Finally, if $i = 0$, that is if there is indifference between present and future, the optimal biomass X_* will be less than X_{msy} for which $G'(X) = 0$. This difference will be more significant if the marginal stock effect is high, therefore with a high unit cost c_y and/or a lesser price p. It is demonstrated in this case that the optimum X_* corresponds to the biomass which maximizes the current economic rentX_{mey} [CLA 85].

The economic approach of the natural capital developed here, which may seem elegant from a mathematical point of view, remains however very theoretical and far from the expected perspectives of application of fisheries economics in support for management. Nevertheless, it has the great merit of explaining the rational choices of renewable resource managers according to their respective preference for the present. This leads to a better understanding of the nature of the trade-offs between conservation and exploitation, which are very acute when decision-makers are confronted with the necessity of fulfilling immediate basic needs (case of many Southern poor countries).

However, this approach is not without theoretical weaknesses. The idea of a choice between investments in the natural resource or in "the bank" refers to the idea of a perfectly malleable capital. Nonetheless, this is not the case with natural resources which, once converted beyond an irreversible threshold into financial capital, will be definitively lost (financial capital can be destroyed and rebuilt, this is unfortunately impossible for natural capital). The natural capital has therefore fundamental attributes (including its bequest and existence values) which are not taken into account by this theory. Here, we again find the elements of the controversy between the supporters of the weak sustainable development [SOL 86, HAR 77] who consider quite legitimate converting, without limits, the natural capital into economic capital and those who, on the contrary, disallow any decrease in natural capital below a critical level [FAU 95], or even any discounting of future incomes [DAL 99], to the detriment of the interests of our descendants. Finally, some other internal criticisms may be addressed to this model. Some conclusions of the model may be reconsidered if the hypothesis of independence between discount rate i and unit cost c_y is abandoned. The discount rate is linked to the interest rate in the national economy. The interest rate positively influences the cost of fixed and variable inputs of the fishery and therefore the unit cost of fishing ($c'_y(i)>0$), which leads, for a higher discount rate, to a reduction in the marginal stock effect.

3.2.4. *The age-structured bioeconomic models*

Most of the bioeconomics models, such as those exposed until now, are based on a very simple representation, even simplistic, of the resource. This limits their ability to take into account the impact of fishing on important characteristics of the exploited natural capital (including its demographic structure), as well as to simulate management measures (such as those concerning the selectivity of fishing equipment or even limiting the effect of fishing on particular components of the natural capital). We are going to present here a family of bioeconomics models whose biological component is based on the pioneering work of Beverton and Holt [BEV 57]. We will present this model in the case of the exploitation of a single cohort, and later we will generalize the results to the case of multiple cohorts.

3.2.4.1. *Bioeconomic model of a simple cohort exploitation*

In this model, we follow the evolution of a set of individuals belonging to the same cohort since its entry (recruitment) in the fishery up to its total depletion. It is considered that the variation of the number of individuals N in the cohort at time t is a function of the total mortality Z_t. Total mortality is the sum of the natural mortality M_t

and the fishing mortality F_t. R is the initial number of individuals recruited[28] in the cohort ($N_0 = R$):

$$Z_t = F_t + M_t \qquad [3.31]$$

It is considered that fishing mortality is applied from $t = t_c$. t_c is a very important parameter of fisheries management, linked to the selectivity of fishing equipment, on which managers may act by modifying the mesh or other technical characteristics. t_c can be related to the age of fish. If fish are recruited at an age a_r, then fishing mortality will begin at age a_r+t_c. From the previous equation, the cohort number in $t+1$ can be expressed as a function of the cohort number and fishing mortality in t:

$$N_{t+1} = N_t e^{-Z_t} \qquad [3.32]$$

The total mortality in number between t and $t+1$ is equal to the difference between N_t and N_{t+1}:

$$N_t - N_{t+1} = N_t.\left(1 - e^{-Z_t}\right) \qquad [3.33]$$

The number of dead individuals due to fishing is calculated by multiplying equation [3.33] by the ratio of the fishing mortality rate and total mortality:

$$\left(\frac{F_t}{Z_t}\right).N_t.\left(1 - e^{-Z_t}\right) \qquad [3.34]$$

The cohort number and the number of caught individuals are known from now on. We now need to compute the same variables with respect to weight. In order to achieve this, an individual growth

28 The recruitment is constituted by the arrival of new individuals likely to be caught in the fishing areas. The recruit number must be distinguished from that of individuals actually caught, which will also depend on the regulatory measures concerning the selectivity of fishing equipment, as well as the seasonal and spatial distribution of the fishing effort (see also *Ecosystem Sustainability and Global Change* [MON 14], Chapter 4).

law is used, which provides the individual weight, w_t. The most commonly used growth function with regard to length is the Von Bertalanffy growth function [BER 38], which seems well suited to fish; other laws can also be considered. The weight w_t is then obtained by means of a weight–length relationship specific to each species. From an economic point of view, it may also be wise to estimate the individual growth in value, especially when price increases with size. This is the case of not only shrimps, but also octopuses and tuna [MUL 11]. In the case of other species, an inverse or nonlinear relationship is observed[29]. The individual value v_t is equal to the product of the individual weight w_t by a price depending on individual weight, or $v_t = w_t.p(w_t)$.

Given the weight and the individual value of an individual of age t, it is possible, by using equations [3.31]–[3.34], to compute the evolution of cohort and catches in weight and value. The biomasses in weight B and value Bv (biovalue) are obtained from equation [3.32]:

$$B_t = N_t.w_t \quad Bv_t = N_t.v_t \qquad [3.35]$$

The catch weight y and instantaneous value yv are calculated by using equation [3.34]:

$$y_t = \left(\frac{F_t}{Z_t}\right).N_t.\left(1-e^{-Z_t}\right).w_t \text{ and } yv_t = \left(\frac{F_t}{Z_t}\right).N_t.\left(1-e^{-Z_t}\right).v_t \quad [3.36]$$

The total catch in weight Y and value Y_v carried out over the lifespan of the cohort is obtained by summing over time the catches at each period:

$$Y = \sum y_t \quad Yv = \sum yv_t \qquad [3.37]$$

Finally, the initial biological version of the model, which aims to determine the conditions of maximization of the total catch with respect to the cohort, proposes as an indicator the total yield per recruit that can be expressed in volume *(Y/R)* or value *(Yv/R)*.

29 This is especially the case for sardines intended for preserving.

Figure 3.22 shows the evolution of a cohort with and without exploitation. Exploitation starts at $t = t_c$ and fishing mortality remains constant until the exhaustion of the cohort. The effect of t fishing mortality is first visible on the curve of the cohort number, then on that of biomasses in volume and in value.

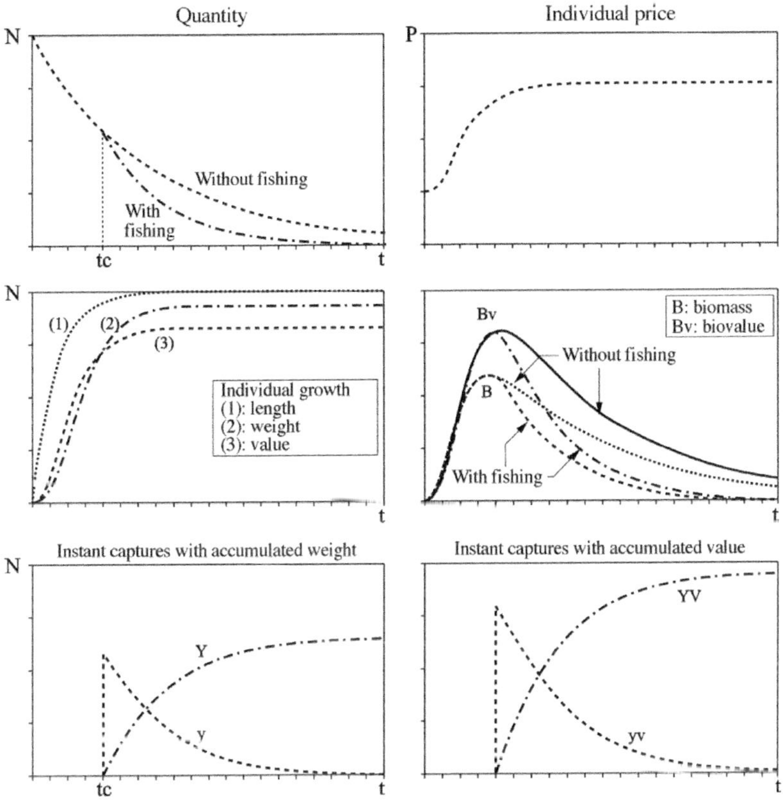

Figure 3.22. *Evolution of an exploited cohort: numbers, biomass in volume and value, instantaneous catches in volume and value*

From an economic point of view, it is important to emphasize here the impact of the variation in price p according to individual weight and then age. It explains notably the time lag observed between the maximum values of two curves representing biomass in value and volume. Without exploitation, the two curves show first a growth phase during which the rapid individual growth outweighs the reduction of the cohort fish number by mortality. The two curves then pass through a maximum beyond which the effect of the mortality Z outweighs that of growth. There is a gap between the two curves because the compensation of mortality by growth lasts longer in value due to the price increasing with age.

This model aims to determine the value of the fishing mortality that maximizes the "objective function" of the manager. Here, the bioeconomic approach diverges from that of the strict biologist because in a similar way to Gordon–Schaefer's model, it is the maximum economic rent (defined as previously) and not that of the catch that must be achieved. The rent obtained from cohort exploitation, in a similar way to the catch, may be reported to the number of recruits.

In order to determine the optimum value of fishing mortality, the impact of the increase of F and t_c on landings volume and value must first be studied.

In order to achieve this, a multiplier of the initial fishing mortality for different values of t_c is applied, and a family of curves is obtained, which are presented in Figure 3.23.

In the simple model presented here, it is considered that the total exploitation cost CT is proportional to the fishing mortality F, which implies that F is proportional to the fishing effort and therefore catchability q is constant whatever the age:

$$CT = c.F \tag{3.38}$$

Figure 3.23 shows that the yield per recruit is relatively sensitive to the increase of t_c (expressed here in months); it is maximal for a value of t_c equal to 3. For low values of t_c (less than 3), any increase in F is accompanied by a decrease of Y/R because an intensive and precocious exploitation of the cohort no longer allows taking advantage of individual growth. On the other hand, for values of t_c equal to or greater than 3, any increase in F, within the range of values presented here, induces a growth in the yield per recruit.

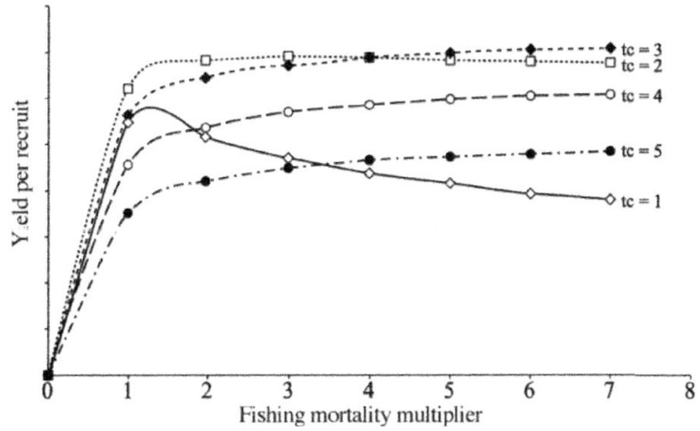

Figure 3.23. *Response of the yield per recruit* Y/R *for different values of* t_c *and of fishing effort*

Figure 3.24 shows the response of the economic rent per recruit to the same parameters; the results are significantly different from those obtained for the yield per recruit. The maximal value of the rent by recruit is also reached for t_c = 3. In contrast, the response to the increase in F is very different; all the curves strongly decrease beyond a value of the multiplier of the fishing mortality equal to 1.5. The conclusion of the bioeconomic model presented here appears, as in Gordon–Schaefer's model, significantly more conservative than that of the biological model.

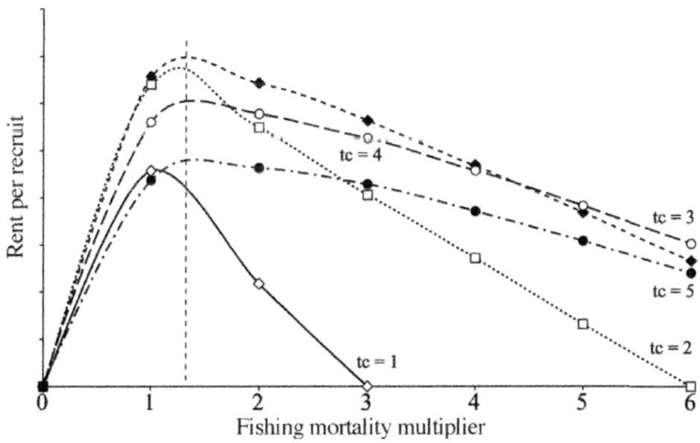

Figure 3.24. *Response of the economic rent per recruit for different values of* t_c *and of the fishing effort*

3.2.4.2. *Extension of the model structured by age in the case of several cohorts*

It is obvious that an exploited population is always composed of several age classes simultaneously exploited. The previous model can perfectly take into account the exploitation of several monthly cohorts within an annual time period by combining them within a single one, whose total recruitment will be equal to the sum of the monthly cohorts. This simplification, often referred to as a model of "pseudo-cohort", is perfectly acceptable if a constant fishing mortality F is applied to all monthly cohorts.

This approach is no longer valid if the fishing effort (and then the fishing mortality) varies during the year. It is therefore not adapted to simulate fishery policies with seasonal closing aiming, for example, to protect the most productive cohorts in their juvenile phase and then achieve a higher commercial value at catch time. In this case, a multi-cohorts model should be used. An example is provided by a bioeconomic model applied to the exploitation of tropical penaeid shrimps [CHA 08a, CHA 11]. In this model, each monthly cohort is modeled separately. Biomass and instantaneous catches are calculated by summing over all the cohorts. Another advantage of this type of model is that it represents more explicitly the effect of selectivity measures on age structure of stock and catches. A selectivity curve can be used to compute at each period a vector of mortality per age which will be applied to different cohorts.

Figure 3.25 shows an application of this type of model to a multi-fleet octopus fishery (artisanal fishing units, coastal and freezer trawlers) with a 3 year simulation period. The fishery is managed by means of two seasonal closures per year.

The upper graph in the figure shows the dynamics of the number of individuals in a few cohorts (all have not been represented for the sake of clarity) and the next graph represents fishing mortality F applied on these cohorts. F increases with the aging of each cohort due to the effect of selectivity in the case of trawlers. During the closed seasons, the fishing mortality is zero, which implies a lesser decrease in the number of individuals of the cohorts. The following graph presents the evolution of the stock. It can be seen that biomass reacts very quickly to the stopping of fishing activities, which is explained by the rapid growth of the species considered.

The last two graphs in Figure 3.25 show the results of economic exploitation: revenue, cost and profit (here equal to rent) accumulated over each year. The observed plateaus are explained by the closure periods. It should be noted that during closures, profit tends slightly to decrease because some costs continue to be incurred (the concern is notably about fixed costs such as depreciation and insurance costs, which are considered, in this model, as distributed over every day of the year).

The last point to mention with this type of model concerns the possibility and benefit of integrating a stock-recruitment relationship (S/R). The recruitment is now made endogenous by considering a relationship, apparently obvious, between parental biomass and its descendance (recruits). It is known that such a relationship is implicit in Gordon–Schaefer's model (see equations [3.1] and [3.2]). In the type of model in which we are interested here, two major types of relationship are usually proposed (Figure 3.26). The Ricker relationship is often chosen when the adults exert a predation on their juveniles, which implies a decrease in recruitment beyond a certain level of spawning biomass, unlike that of Beverton and Holt's where the relationship is always increasing and tends toward a horizontal asymptote.

The integration of a stock recruitment relationship allows us to build models of multiannual simulation to represent in a more realistic manner the long-term consequences of management policies. The application of this approach is, however, limited by the great difficulty in making empirical estimates of S/R relationships such as those presented in Figure 3.26. For many species, the impact of natural environment looks more decisive for the success of recruitment than the abundance of parental stock. For some species such as penaeid shrimps, the majority of authors even consider that there is no stock recruitment relationship at all. On the other hand, most of the authors recognize that there is, regardless of the species, a critical fecund biomass value below which the recruitment tends to decrease.

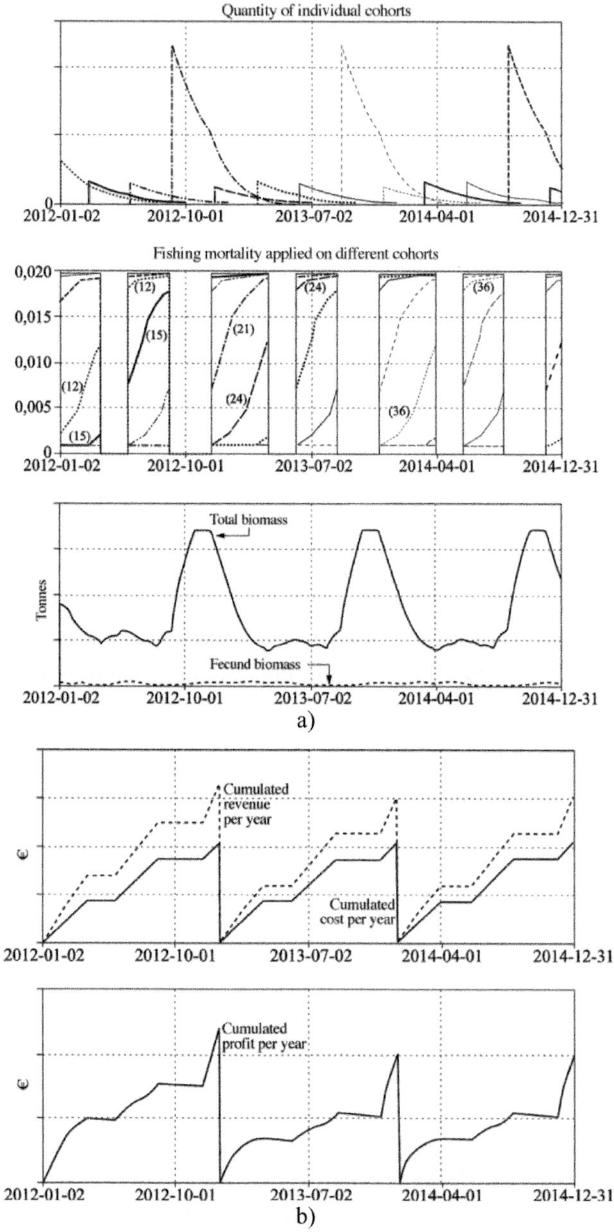

Figure 3.25. *Age structured bioeconomic modeling of a cephalopod fishery*

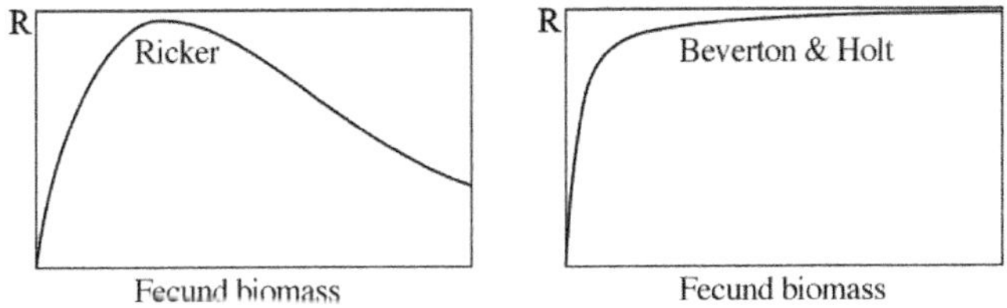

Figure 3.26. *Ricker, and Beverton and Holt stock recruitment relationships*

In the absence of an S/R relationship, the retrospective analysis of effective catches by Virtual Population Analysis[30] may allow an estimate of past recruitments and of their variability. Stochastic bioeconomics simulation models can thus be proposed, which integrate the effect of the natural environment on recruitment and therefore on the economic results of fisheries [CHA 08b].

To conclude with the age-structured bioeconomic models, it should be mentioned that they are particularly suitable for simulating management measures that rely on seasonality and the age structure of the natural capital. The conclusions on economic overexploitation are identical to that of Gordon–Schaefer's model. On the other hand, the biological approach to overfishing is more complex here since two distinct phenomena are taken into account. The first is the growth overexploitation, when fishing effort is exerted on individuals that are too young and the majority of which will not reach their maximal market value. The second is recruitment overexploitation, when the decrease in fecund biomass implies a subsequent decrease in recruitment. Although essentially biological, these phenomena are not to be ignored by economists, first because they affect the long-term sustainability of the main factor of production and then also impact economic results in the short, medium and long terms. This type of model, however, needs more expensive information, notably regarding

30 The Virtual Population Analysis (VPA) is a method that allows through the analysis of the effective catches structure (by size, or by age) the estimation of the number of individuals in cohorts and, therefore, their initial recruitment. It uses the same equations as the previous model, but instead of following the cohort individual numbers over time, they are estimated by retro-computations from the elder individuals, and by formulating a hypothesis on the final value of fishing mortality. We refer the interested readers to the chapter dealing with this method in more detail.

the biological components, and its results appear to be very sensitive to the quality of the available biological parameters (notably individual growth and natural mortality).

3.2.5. *Toward more realistic and complex bioeconomic models*

In the context of this contribution, it is not possible to present the whole diversity of bioeconomic models. However, before concluding about bioeconomic modeling, it is important for us to make reference to a few approaches developed over the past two decades and which are extensions and further investigations of the more general approaches that have been presented so far.

3.2.5.1. *Spatialized models*

Traditional bioeconomic modeling does not explicitly take space into account. This limits its ability to account for spatial structures which are important both from the resource and the exploitation points of view. The spatial distribution of certain species can thus vary during their lifecycle, which is the case, for example, of penaeid shrimp species whose juvenile and subadult phases are coastal (and therefore accessible to techniques of artisanal fisheries) and the adults are present further away offshore and exploited by trawlers [CHA 08a]). Spatialization also allows us to take into account the impact of the distance between landing places and fishing grounds on the exploiting costs. Finally, a spatialized model is obviously more appropriate for the simulation of measures, such as marine protected areas (MPAs). A general presentation of spatialized bioeconomics models is proposed by Anderson and Seijo [AND 10]. This ambitious approach has resulted in generic applications such as the ISIS-FISH model [MAH 04, PEL 09], which was used to simulate the responses of many fisheries under various management measures, such as MPAs among others.

3.2.5.2. *A finer representation of fishing strategies*

The bioeconomics models presented so far also appear relatively simplistic in the representation of the choices of fishermen, collected through the only relationship between profit and fishing effort.

However, other decisions should be taken into account, such as those concerning the choices of fishing areas and target species in artisanal fisheries characterized by a great practice of diversity. Such a model has been proposed for the Senegalese artisanal fishery [LAL 07, PEC 01]. Results show the redeployment of fleets (fisheries strategies) between the different available fishing tactics, the so-called "métiers", according to their present and past economic benefits (a year earlier). This model is one of the few applications of McFadden's approach of discrete choices [MCF 73] to fisheries.

3.2.5.3. *Viability of bioeconomic modeling of fisheries*

The bioeconomics models remain largely dominated by the monocriterion optimization approach (the latter being almost always the economic rent). The application of the viability theory in fisheries [LAR 08] allows new directions and perspectives in bioeconomic modeling. The main point is to maintain the fishing system in a viable multidimensional space that simultaneously respects several constraints. These constraints concern, for example, the biological resource, the income of fishermen and food provisioning of households [HAR 13]. The maintenance within the viable space may be allowed by loops of endogenous feedbacks (rules of behavior) or by exogenous control variables (in a similar manner to the more conventional bioeconomic models). Some viability models allow that the trajectory moves temporarily out of the area of viability (crisis). Their approach aims to find out the conditions for the minimization of the crisis time (this is the time during which the trajectory moves away from the viable space). Contrary to what is sometimes said, the viability theory does not therefore reject optimization, but uses it in a different way.

3.3. Contribution of economics to fisheries management

3.3.1. *A negative view about the efficiency of public policies in the fishery sector*

The management of fisheries concerns all the tools implemented by the public policies concerning the field of fisheries. It is possible to say without any risk that these public policies are far from having achieved their objectives and that the fishing world is in crisis. First, this is due to the critical state of a large number of fishing stocks and the consequences of fishing activities on marine ecosystems [CUR 04]. Seen from a more economical perspective, the policies followed have often been expensive for society without achieving their expected objectives. A study from the World Bank [WOR 08] proposed an economic evaluation of worldwide fisheries in 2004. It compares the actual economic results of worldwide fisheries to the profits that would have been obtained from an optimal exploitation. Two models, using slightly different biological assumptions (Schaefer's and Fox's models), have been used. They led to relatively similar conclusions.

	Current situation		Optimal situation		Difference	
	Model 1	*Model 2*	*Model 1*	*Model 2*	*Model 1*	*Model 2*
Biomass (10^6 t)	148.4	92.3	314.2	262.9	165.8	170.6
Catches (10^6 t)	85.7	85.7	8.8	81.6	-4.9	-4.1
Index of effort	100	100	0.56	0.46	-0.44	-0.54
Economic rent (10^9 US $)	-5	-5	39.5	54	44.5	59

Model 1: Schaefer; Model 2: Fox

Table 3.5. *Comparison of current situations and optimal worldwide fisheries by means of two model (source: [WOR 8])*

The estimated loss is in the order of 50 billion dollars. The optimal situation, in terms of economic rent, would be obtained for reductions of 5% of the catches and 50% of fishing effort. Depending on the chosen biological model, fish biomass would double or triple with an optimal exploitation. Extrapolated to the past three decades (from 1974 to 2007) and assuming that the rent has followed a similar trend to that of the state of world stocks according to the FAO, the economic loss is estimated at US $2.2 trillion. This result could be assigned to the low performance of the management of fisheries in Southern countries where the research and regulation policies are inadequate. However, the analysis of national experiences shows that rich countries are far from being good examples or even better performing. A recent study about the French case [MES 08] highlighted astonishing figures: public aids granted to the fisheries (including the support of the state to the social security of fishermen) over the four previous decades would approximate the value of landings. In spite of this considerable effort by the whole society, the pursued objectives have not been achieved, both in terms of profitability of the sector and the status of exploited resources.

These figures highlight the necessity for a management that should be at the same time more rational (in a purely economic sense) and sustainable (taking into account the rights of future generations, as well as the three ecological, economic and social dimensions of sustainability). One of the causes for weak economic performances would be the inadequacy of public policies in the fisheries management. Despite the recommendations from bioeconomics literature since the 1950s, which explain the sources of biological (when fishing beyond the maximum of balanced production) and economic overexploitation (when the means invested in the exploitation of the resource are higher than those allowing the maximum profitability of the exploitation), fisheries management policies have often been failed. For political reasons, but also due to the deficiency of ideas at the origin of public action, fisheries management has long supported the development of the fishing activity without taking into account the state of the exploited natural resources and the poor economic performance with regard to the national interests. Among others, subsidy policies helping the development of fishing capacities, aimed at increasing food self-sufficiency or even supporting the competitiveness in a context of increasing globalization, have been the subject of increasingly more critical analyses, calling for a drastic reorientation of public action. The most recent estimate considered that the amount of subsidies was between 25 and 29 billion dollars in 2003 [SUM 10], and accounted for approximately 35% of the value of worldwide landings.

3.3.2. *A classification attempt of the tools for fisheries management*

There is a large set of possible fisheries management measures, from the most technical to those based on elaborate economic mechanisms such as fishing rights markets. These different management tools can be distinguished according to a double classification.

The first classification between possible measures is obtained according to the variables that are subject to regulation. Therefore, the measures controlling the inputs will be distinguished from those regulating the production (output), as well as those allowing the control of space or the seasons open to fishing. The second classification is based on the modalities of implementation: regulatory measures can therefore be distinguished from those based on monetary incentives (taxes/subsidies) and from those based on rights markets. We will not cover here in detail the measures presented in Table 3.6; we refer to the chapter dealing more specifically with the management of the fisheries. We will instead discuss the economic approach which allows its benefits to be assesed.

<table>
<tr><th rowspan="2"></th><th colspan="2">Control of the means of production</th><th rowspan="2">Control of production</th><th rowspan="2">Spatiotemporal measures</th></tr>
<tr><th>Direct</th><th>Indirect</th></tr>
<tr><td>Regulatory</td><td>Licenses
Permits
Non-tradable equipment quotas</td><td>Equipment and boats characteristics (length, power)</td><td>Total allowable catch (TAC)
Legal sizes of fish caught</td><td>Marine protected areas
Seasonal closures</td></tr>
<tr><td>Monetary incentives</td><td>Fishing royalties</td><td>Taxes or subsidies on/for inputs (fuel, equipment, etc.)
Subsidies for fleet decommissioning</td><td>Taxes or support on/to ex vessel prices</td><td></td></tr>
<tr><td>Rights market</td><td>Licenses, permits, transferable quotas for equipment</td><td></td><td>Individual transferable quotas (ITQ)</td><td></td></tr>
</table>

Table 3.6. *A possible classification of the main measures regarding fisheries management*

3.3.3. *An economic assessment of the relevance of the tools of fisheries management*

Different criteria should be used in order to judge the utility and relative effectiveness of the different measures:

– the ability to achieve the targeted objectives. The latter are very diverse: restoration of the activity profitability (which can be considered at the collective – rent – or private level – profit of the firm), resource conservation, fishermen incomes, employment, food security, etc., it appears that these multiple goals are often concurrent [CHA 01]. It is wise, for the clarity of the presentation, to limit here our questioning to the first two;

– the costs of implementation of measures, of the necessary information and of the coordination between the actors (transaction costs);

– the degree of acceptability by all actors active in the fishery;

– the strategic behavior of fishery actors in response to public policies. Fishermen should not be regarded as passive when faced with the implementation of management measures.

3.3.3.1. *The control of the effort and the capacity of fishing*

The system of permits or licenses for fishing is the most generalized management mode. These licenses can be free or not (royalties from fishing). The instauration of licenses is also an obligation of the flag state for units operating on the high seas. This minimal system is, however, not always systematically put in place or respected in Southern countries. For example, traditional Malagasy fishery is not yet subject to this type of control; in Senegal, a recent project of registration of pirogues has enabled progression in this direction, but its application is not yet widespread. Very often, it is a simple registration system which does not define restricting limits to the fishing capability.

To be efficient, such a system must be restrictive and must impose explicit limits to the entry in the fishery. Another difficulty is the existence of authorized but not very active vessels, which represent a potential fishing capacity when the fishery will become more profitable. The licensing system should not only limit the number of units but also their characteristics (most often tonnage and power), in order to prevent the transfer of the permits toward more powerful units, during the renewal of fishing units. When the licenses are transferable between fishermen, they can reach a very high value and any attempt to reduce the fleet may be expensive for the state. Then it may be wise to limit the transfer possibilities, or to limit the validity of the licenses in time. The French experience shows, however, that the legal limitation of the transferability of fishing permits can be overcome. Although not legally transferable, the value of the exploitation permits should represent 50% of the market price of second-hand fishing units [BUI 07].

When a fishery is being overexploited, the simple system of fishing licenses is insufficient because the aim is then to reduce the fishing capacity. This reduction can be achieved in three ways: removing some boats from the fleet, reducing the authorized fishing time per boat, and restricting the number and size of fishing equipment per boat. In addition to the fact that any withdrawal must give rise to financial compensation, these measures induce effects which limit their efficiency (the least efficient units are removed in priority and the reduction in capacity is then not proportional to that of the fleet, and the monetary compensations may be reinvested to increase the productivity of remaining units). Finally, there is a risk of exporting the fishing capacity to third-world countries (well-known case of the second life of some ex-European boats in West Africa). The reduction of the number of days at sea, or fishing hours, is a fairly straightforward measure, particularly used in the Mediterranean by the professional organizations to which a part of the management is delegated. Finally, the fishing time regulation can be seasonal closures, also called "biological rest", aimed at both a reduction of fishing time and resource conservation. They are particularly used for short lifecycle species and/or characterized by a strong seasonal variability (shrimp, octopus, etc.). The measures relating to the season or to fishing time are fairly simple to implement. However, they can be easily circumvented by intensifying the use of non-controlled fishing means when fishing is allowed. Management by "biological rest" can then encourage a frantic race for the resource (fisheries qualified as "Derby Fisheries" [CLA 02]) and high exploiting costs undermining the profitability of boats. Limiting fishing hours per day will lead to a strategic response by fishers through the use of a greater quantity or size of equipment, if not controlled.

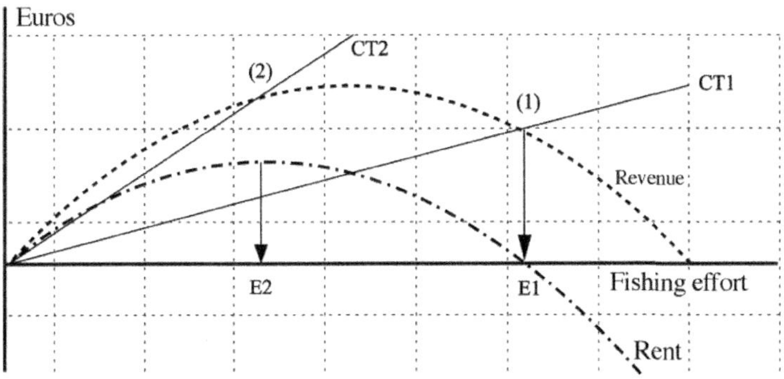

Figure 3.27. *Use of a bioeconomics model to expose the consequences of different reduction measures of the fishing effort*

The choice between the different means to control fishing effort has major consequences in terms of distribution, or even equity, and will then affect the social acceptability of the regulation put in place. This issue is illustrated in Figure 3.27, where three solutions are possible to move the fishery from a free access equilibrium E1 to the bioeconomic optimum E2: (1) reducing the number of vessels from E1 to E2 by reducing the number of licenses, (2) taxing the fishing effort (royalties) in order to shift total cost from CT1 to CT2 and obtain a new free access equilibrium in E2, and (3) a taxing effort (royalties) and limiting the number of licenses at the same time. In case 1, the maintained fishing units monopolize the rent; in case 2, the state gets the rent through the amount of levied taxes; and finally solution 3 allows a sharing of the rent between the state and the maintained fishing units. Solutions 2 and 3 allow the state to financially help the reconversion of fishers leaving the fishery by allocating part of the rent to this aim.

3.3.3.2. *Catch control*

Three major types of measure can be distinguished here:

– the limitation of total catches by TAC (total allowable catches) whose main objective is to reduce or maintain catch at a sustainable level (MSY);

– quotas, obtained by the distribution of the total allowable catch among fishermen, groups of fishermen or even countries (in the case of international fisheries). These quotas can be transferable through market transactions. They are then considered by many authors as the most effective tool in the management of fisheries (because they are based on a sound economic incentive principle) but nevertheless remain the subject of criticism (concentration of fishing rights, financialization of the sector, difficulties of application, resource privatization, etc.);

– the control of the size of the fish caught aims to reduce the catch of immature individuals or low commercial value through an action on the selectivity of fishing gear, that is to say on their ability to capture individuals beyond a certain size (mesh size of trawls, gillnets, size of hooks, etc.). The establishment of a minimum size of fish caught can complement the previous measure.

Although generally preferred to the control of production means [BUI 07], catch control implementation is also faced with serious difficulties. The first difficulty is the definition of the TAC, which must be re-estimated each year due to the variability of the resource. This assumes access to good, but expensive catch statistics. These conditions restrict in consequence catch control to high economic value resources. This approach is therefore very difficult to apply in multispecies fisheries such as the majority of artisanal fisheries of the Southern countries or even Mediterranean fisheries[31].

31 With the exception of bluefin tuna for which a quota per country is defined, following the decisions of the ICCAT. It is then redistributed to the national segments according to the rules of each country, without transferability.

Used alone, the TAC can encourage fleet overcapacity (race for fish or *derby fishery*): each fisherman fishing the maximum before the TAC is reached, which leads to significant costs and therefore poor economic results. The resulting peaks in landings are difficult to absorb by onshore industries (canning, fishmeal and cold storage facilities) which are also encouraged to overcapacity. When TAC is distributed in the form of quotas to different sectors or producer organizations (POs), it is difficult to obtain reliable catch statistics at their level. The risk of a fishing race still exists within the sectors or the POs, the different members being in competition for sharing the collective quota.

The allocation of the TAC in quotas to vessels or individuals is therefore supposed to be more efficient because everyone uses their available production means in the best economic way in order to capture his quota. However, some difficulties still remain: good reliable catch statistics at the individual level are necessary (risk of under-declaration). One of the most important questions here is the incentive to discard species or sizes with little market value. When the individual or collective quota is reached for a species, any additional catch of this species must be discarded. This issue is a stumbling block for the current reform of the European Common Fisheries Policy which aims at both an enlargement of individual quotas and the stopping of discards at sea. This contradiction may well be resolved by the creation of a legal market to valorize over-quota catches previously discarded. However, this solution may encourage the fishers to develop their activity strategically according to this new opportunity.

If individual quotas are transferable, there may be an opportunity for the most efficient fishermen to concentrate individual fishing rights. This may be positive from a strictly economic point of view, but can also have adverse social consequences for the maintenance of the cohesion of fishing communities, and then work against the societal acceptability of this tool. There is finally a risk of dissociation between the ownership of fishing rights and the practice of fishing, the fishermen then working on behalf of the owners of the catch property rights.

The last-mentioned measure, management by size limit, is not without difficulties: the selectivity of fishing gears is sometimes limited, and there is a significant risk of small fish discards. Fish discards are inevitable for all catch with illegal size.

3.3.3.3. *Marine protected areas*

Faced with the difficulties of the conventional fisheries management instruments, scientists have promoted MPAs as an effective means of reconciling the conservation and exploitation of fishery resources [CLA 11]. According to the International Union for the Conservation of Nature (IUCN), PMAs are "any intertidal or subtidal space, with the volume of water that covers it, the fauna, the flora, the historical and cultural characteristics associated, which is placed in reserve by law or by other means to protect all or part of the environment therefore enclosed". Following the world congress on sustainable development in 2002, the international community has agreed for a commitment to build a global network of MPAs. France, for instance, aims in the long term at the classification of 20% of its EEZ in MPAs. In order to achieve it, it has created the marine protected areas agency (*Agence des Aires Marines Protégées*) in 2006 [CHA 11].

Unlike the conventional fisheries tools discussed previously, PMAs aim at a broader set of objectives, such as the conservation and the protection of marine resources (whether commercial or not) and their habitats, the protection of endangered or emblematic species and their habitats, and the protection of unique habitats. It further includes the protection of high biodiversity areas or high biological productivity, the sustainable management of fisheries and coastal areas and coastlines.

Although fishing is only one of the multiple aspects taken into account in the public policies of PMAs, it is important to emphasize their main contributions to fisheries management. The concern is initially about the effects on the abundance of the natural capital. If the MPA is well designed, it is expected to export after a few years (spillover effect) fish at larval and juvenile, or even adult, stages, out of the MPA. It can also be conceived to limit the catches during certain critical phases of the fish lifecycle. This is the case of permanent or temporary MPAs closing the breeding areas of tropical groupers. It can also allow the protection of critical habitats of the lifecycle of commercial species (e.g. mangroves, sea grass beds and coral reefs).

MPAs rely on a spatial planning of all the activities in the concerned area, and then allow limiting multiple use conflicts, that are quite frequent in coastal areas. Finally, in developing countries, MPAs are a means to protect the access of traditional communities to resources (for instance, Fiji Islands and New Caledonia) by strengthening the traditional rights jeopardized by history and the development of industrial or recreational concurrent fisheries.

During recent years, many works in bioeconomics have analyzed the conditions under which MPAs could contribute to the durability of fisheries; we will not expose these works here because they are in future book in the "Seas and Oceans" set. Nevertheless, it appears important to rapidly discuss the limits of PMAs, as we have done for the other tools of fisheries management.

The question about the impact of the resource effect on the profitability of fishing activities can be raised first of all. If it is admitted that the majority of MPAs produce a positive effect on the abundance of natural capital, it will lead to better economic results for the fishermen only if the export of resources from the non-fishing era is greater than the quantity of resources lost for fishermen after MPA creation.

The strategic reactions of fishers to MPA creation can counterbalance the positive resource effects. A well-known case is the concentration of the fishing effort to limit the protection zone (often designated by the term *fishing the line*). This concentration of fishing effort may be due to the fishermen initially present, but it may also be the result of outside fishermen attracted by the reserve effect. As any phenomenon of congestion, this concentration generates negative externalities, which negatively impact the fishery economic performance. As a consequence, local fishermen may intensify their fishing effort due to the increased competition for resource around the MPA, with negative consequences on fishing costs.

An MPA policy will not solve alone the major problem of fishing, which is the chronic fishery overcapacity. In the absence of control capacity, MPA creation may lead to a constant or growing concentrated fishing effort on a more restricted area.

Finally, MPA is characterized by immediate and certain costs for the fishermen (opportunity cost equal to the loss in catches formerly made in the closed area) and distant and uncertain future gains. The revenue uncertainty is both due to the reserve effect and due to the diffuse nature of the profits that are shared with other professional fishermen or even other marine resources users (recreational fisheries among others). Compliance to MPA policies to MPA from fishermen unless some concessions are made to earn their opinion. This leads, with an increasing role of MPA policies with the expansion of national networks, to move away from the initial objective of conservation (model of the marine reserve) to that of concerted management of a multi-use marine space (model of the marine park), which is less restrictive for fishermen. It seems then too early to judge whether this evolution does confirm or not the effectiveness of MPAs regarding the sustainable management of fisheries.

3.4. Conclusion: the contributions of fisheries economics and its future evolution

In 1954, Scott Gordon did not hesitate to write that fisheries management should be done on economic basis and not on biological one and that "the ultimate question is not the ecology of life in the sea as such, but man's use of these resources for his own (economic) purposes" [GOR 54]. Today, these words may seem from a different age, whereas there is an attempt to impose the ecosystemic approach to fisheries as a general frame of public policies. However, the understanding of individual and collective human responses to public policies and to economic incentives still remains necessary and important, and creates a continuous need for economic science. The prospects and societal expectations relating to marine natural resources have today expanded, and it has consequences for the scientific agenda. The social utility of natural resources can no longer only be limited to the rent resulting from exploitation but should be extended to other components of the total value of the natural capital which integrates use and non-use values. Among the latter, it is necessary to retain the existence value [GRO 02], which should allow better integration the societal objectives of conservation.

The concern is here about a first possible direction for the economics of marine natural resources (and no longer only fisheries economics), more closely to environment economics which has produced many different methods for the assessment of the natural market and non-market assets [JOH 99]. In this perspective, future works should be directed toward an assessment of the true costs of fishing, taking into account the negative externalities of fishing on the whole ecosystemic marine services. This is a very ambitious program which raises many conceptual and methodological problems. Progression in this direction is possible from the existing theories and scientific knowledge of fisheries and environment economics, and of a new rapprochement with ecological sciences, essential for the understanding of the ecological functions underlying the ecosystemic marine services.

Over the past few years, the economic dimension has been explicitly incorporated into complex ecosystemic models such as *Atlantis* [FUL 11] and *EcosimWithEcopath* [CHR 04]. In *Atlantis*, a fleet dynamics model is included where costs and revenues are computed either on the basis of enforced decisions or on an endogenous response of the fleets to their past and present profits. *EcosimWithEcopath* incorporates a model of economic evaluation of management choices by means of a collective utility function, including the economic rent, the existence value of the resources and the uncertainty associated with the different decisions. These two cases are interesting examples of the integration of economic components based on conventional scientific approaches. Comprehensive applications of these models remain, however, relatively scarce to this day, due to the amount and cost of information necessary to supply all of their components.

Finally, future developments should also enable us to go further into the understanding of the microeconomic behavior of the actors and to take better account of their heterogeneity. Most of the models presented here are based on relatively simplistic behavior hypotheses and especially rely on the "fiction" of the representative agents whose limits are known [KIR 92]. The heterogeneity of the agents may be relative to their technical efficiency, to their risk aversion, to the behavior of investment, to the preferences relating to the state of ecosystems and to their degree of individualism (models can be imagined where individuals would not all be strictly selfish but where the utility function would have the well-being of their neighbors as attribute). Such a path of research is possible by agent-based modeling, today well known in principle, but whose use could be usefully developed in the field of marine natural resources economics.

3.5. Bibliography

[ADA 68] ADAM P., "Aspects économiques de la "surpêche"", *Revue Economique*, vol. 19, no. 1, pp. 130–161, 1968.

[ALC 87] ALCHIAN A., "Rent", in EATWELL E., *et al.* (eds.), *The New Palgrave: A Dictionary of Economics*, vol. 4, MacMillan, London, pp. 141–143, 1987.

[AND 77] ANDERSON L.G., *The Economics of Fisheries Management*, The Johns Hopkins University Press, Baltimore, London, 1977.

[AND 83] ANDRO M., DORVAL P., LE BOUAR G., *et al.*, Analyse des données sur les accidents du travail à la pêche maritime dans les pays de la communauté économique européenne de 1977 à 1980, Lorient University Institute of Technology (UBO), Lorient, 1983.

[AND 10] ANDERSON L.G., SEIJO J.C., *Bioeconomics of Fisheries Management*, Wiley-Blackwell, Ames, 2010.

[ARN 07] ARNASON R., "Property rights quality and efficiency of fisheries management regimes: some basic results", in BJORNDAL T., *et al.* (eds.), *Advances in Fisheries Economics*, Blackwell Publishing, Oxford, pp. 32–58, 2007.

[BEN 10] BÉNÉ C., HERSOUG B., ALLISON E.H., "Not by rent alone: analysing the pro-poor functions of small-scale fisheries in developing countries", *Development Policy Review* vol. 28, no. 3, pp. 325–358, 2010.

[BER 01] BERKES F., MAHON R., MCCONNEY P., *et al.*, *Managing Small-Scale Fisheries: Alternative Directions and Methods*, International Development Research Center, Ottawa, 2001.

[BEV 57] BEVERTON R.J., HOLT S.J., "On the dynamics of exploited fish populations", *Fishery. Investigations*, Ser. II, vol. 19, p. 533, 1957.

[BIN 99] BINMORE K., *Jeux et théorie des jeux*, De Boeck University, Paris, 1999.

[BLA 86] BLAUG M., *La pensée économique. Origine et développement*, Economica, Paris, 1986.

[BON 98] BONNIEUX F., DESAIGUES B., *Economie et politiques de l'environnement*, Dalloz, Paris, 1998.

[BUI 07] BUISSON G., BARNLEY M., Les quotas de pêche individuels transférables. Bilan et perspectives pour une gestion durable des ressources, Etudes et Synthèses, Ministry of Ecology and Sustainable Development Planning, Paris, 2007.

[CAV 08] CAVERIVIÈRE A., CHABOUD C., RAFALIMANANA T. (eds.), *Les crevettes côtières de Madagascar. Biologie exploitation, gestion*, IRD, Paris, 2008.

[CHA 01] CHARLES A., *Sustainable Fishery Systems*, Blackwell Science, Oxford, 2001.

[CHA 07] CHABOUD C., "Gérer et valoriser les ressources marines pour lutter contre la pauvreté", in GOEDEFROIT S., REVÉRET J.P. (eds.), *Quel développement à Madagascar?, Etudes Rurales*, vol. 178, pp. 197–212, 2007.

[CHA 08a] CHABOUD C., "Modélisation bioéconomique de la pêcherie crevettière Malgache", in CAVERIVIÈRE A., *et al.* (eds.), *Les crevettes côtières de Madagascar. Biologie, exploitation, gestion*, IRD, Paris, pp. 261–310, 2008.

[CHA 08b] CHABOUD C., "Une extension pluriannuelle du modèle bioéconomique de la pêcherie crevettière malgache", in CAVERIVIÈRE A., *et al.* (eds.), *Les crevettes côtières de Madagascar. Biologie, exploitation, gestion*, IRD, Paris, pp. 311–326, 2008.

[CHA 11a] CHABOUD C., VENDEVILLE P., "Evaluation of selectivity and by catch mitigation measures using bioeconomic modelling. The cases of Madagascar and French Guiana shrimp fisheries", *Aquatic Living Resources*, vol. 24, no. 2, pp. 137–148, 2011.

[CHA 11b] CHABOUD C., GALLETTI F., DAVID G., *et al.*, "Marine protected areas and governance: towards a multidisciplinary approach", in AUBERTIN C., RODARY E. (eds.), *Protected Areas, Sustainable land?*, Ashgate, Farnham (UK)/IRD, Marseille, pp. 31–49, 2011.

[CHA 89] CHABOUD C., KEBE M., "La distribution en frais du poisson de mer au Sénégal, commerce traditionnel et interventions publiques", *Cahiers des Sciences Humaines*, vol. 25, nos. 1–2, pp. 125–143, 1989.

[CHA 91] CHABOUD C., CHARLES D.E., "Les pêches artisanales en Afrique de l'Ouest: état des connaissances et évolution de la recherche", in DURAND J.R., *et al.* (eds.), *La recherche face à la pêche artisanale. Symposium International Orstom-Ifremer*, Montpellier, France, 3–7 July 1989, IRD, Paris, 1991.

[CHA 95] CHABOUD C., "Risques et incertitudes dans les pêches: le point de vue de l'économiste", in LALOË F., *et al.* (eds.), *Questions sur la dynamique de l'exploitation halieutique, Actes de la table Ronde de Montpellier*, 6–8 September 1993, Coll. Colloques et Séminaires, ORSTOM, Paris, pp. 297–330,1995.

[CHR 04] CHRISTENSEN W., WALTERS C.J., "Ecopath with Ecosim: methods, capabilities and limitations", *Ecological Modeling*, vol. 172, pp. 109–139, 2004.

[CLA 06] CLARK C.W., *The Worldwide Crisis in Fisheries: Economic Models and Human Behavior*, Cambridge University Press, Cambridge, 2006.

[CLA 11] CLAUDET J. (ed.), *Marine Protected Areas: A Multidisciplinary Approach*, Cambridge University Press, Cambridge, 2011.

[CLA 75] CLARK C.W., MUNRO G.R., "The economics of fishing and modern capital theory: a simplified approach", *Journal of Environmental Economics and Management*, vol. 2, pp. 92–106, 1975.

[CLA 79] CLARK C.W., CLARKE F.H., MUNRO G.R., "The optimal exploitation of renewable resource stocks: problems of irreversible investment", *Econometrica*, vol. 47, no. 1, pp. 25–47, 1979.

[CLA 83] CLARK C.W., LAMBERSTON R.H., "An economic history and analysis of pelagic whaling", *Marine Policy*, vol. 6, pp. 103–130, 1983.

[CLA 85] CLARK C.W., *Bioeconomic Modeling and Fisheries Management*, Wiley Interscience, New York, 1985.

[CLA 87] CLARK C.W., "Renewable resources", in EATWELL E., *et al.* (eds.), *The New Palgrave: A Dictionary of Economics*, vol. 4, MacMillan, London, pp. 139–141, 1987.

[CLA 90] CLARK C.W., *Mathematical Bioeconomics: The Optimal Management of Renewable Resources*, Wiley Interscience, New York, 1990.

[CON 10] CONRAD J.M., *Resource Economics*, 2nd ed., Cambridge University Press, New York, 2010.

[CRE 96] CREAN K., SYMES D., *Fisheries Management in Crisis*, Fishing News Books, Oxford, 1996.

[CUR 08] CURY P., MISEREY P., *Une mer sans poisson*, Calmann-Lévy, Paris, 2008.

[DAL 99] DALY H., *Ecological Economics and the Ecology of Economics: Essays in Criticism*, Edward Elgar, Cheltenham, 1999.

[DAS 93] DASGUPTA P., *An Inquiry into Well-Being and Destitution*, Oxford University Press, Oxford, 1993.

[DEG 02] DE GROOT R.S., WILSON M.A., *et al.*, "A typology for the classification, description and valuation of ecosystem functions, goods and services", *Ecological Economics*, vol. 41, pp. 393–408, 2002.

[DES 93] DESAIGUES B., POINT P., *Economie du patrimoine naturel. La valorisation des bénéfices de protection de l'environnement. La valorisation des bénéfices de protection de l'environnement*, Economica, Paris, 1993.

[FAU 95] FAUCHEUX S., NOËL J.F., *Economie des ressources naturelles et de l'environnement*, Armand Colin, Paris, 1995.

[FUL 11] FULTON E.A., LINK J.S., *et al.*, "Lessons in modelling and management of marine ecosystems: the Atlantis experience", *Fish and Fisheries*, vol. 12, no. 2, pp. 171–188, 2011.

[GOR 54] GORDON H.S., "The economic theory of a common-property resource: the fishery", *Journal of Political Economy*, vol. 62, pp. 124–142, 1954.

[GRA 04] GRAFTON R.Q., ADAMOWICZ W., DUPONT D., *et al.*, *The Economics of Environment and Natural Resources*, Blackwell Publishing, Malden, 2004.

[GRA 06] GRAFTON R.Q., KIRCKLEY Y., KOMPAS T., *et al.*, *Economics of Fisheries Management*, Ashgate, Farham, 2006.

[GRO 16] GROTIUS H., *Mare Liberum*, Oxford University Press, New York, 1916.

[GUE 12] GUERRIEN B., *Dictionnaire d'analyse économie*, La Découverte, Paris, 2012.

[HAN 93] HANNESSON R., *Bioeconomic Analysis on Fisheries*, Fishing News Books, Oxford, 1993.

[HAR 13] HARDY P.Y., BÉNÉ C., DOYEN L., *et al.*, "Food security versus environment conservation: a case study of Solomon Islands small-scale fisheries", *Environmental Development*, vol. 8, pp. 38–56, 2013.

[HOT 31] HOTELLING H., "The economics of exhaustible resources", *Journal of Political Economy*, vol. 39, pp. 137–175, 1931.

[HAR 68] HARDIN G., "The tragedy of the commons", *Science*, vol. 162, no. 3859, pp. 1243–1248, 1968.

[HAR 77] HARTWICK J.M., "Intergenerational equity and the investing of rents from exhaustible resources", *The American Economic Review*, vol. 67, no. 5, pp. 972–974, 1977.

[JOH 99] JOHANSSON P.O., "Theory of economic valuation of environmental goods and services", in VAN DEN BERGH J.C.J.M. (ed.), *Handbook of Environmental and Resource Economics*, Edward Elgar Publishing, Northampton, pp. 747–754, 1999.

[KNI 21] KNIGHT F., *Risk, Uncertainty and Profit*, Houghton and Mifflin, Boston, 1921.

[LAL 07] LALOË F., "Information at the interface between supply and demand for indicators: the use of a model with estimated parameter values as a base for indication in multicriteria analyses", *International Journal of Sustainable Development*, vol. 10, nos. 1–2, pp. 33–45, 2007.

[LAR 08] LARA M. DE, DOYEN L., *Sustainable Management of Natural Resources: Mathematical Models and Methods*, Springer, Berlin, 2009.

[KIR 92] KIRMAN A.P., "Whom or what does the representative individual represent?", *Journal of Economic Perspectives*, vol. 6, pp. 117–136, 1992.

[LAU 81] LAUREC A., LE GUEN J.C., *Dynamique des populations marines exploitées, tome 1: concepts et modèles*, CNEXO, Oceanographic Centre of Brittany, Brest, 1981.

[MAH 04] MAHÉVAS S., PELLETIER D., "ISIS-FISH, a generic and spatially-explicit simulation tool for evaluating the impact of management measures on fisheries dynamics", *Ecological Modelling*, vol. 171, pp. 65–84, 2004.

[MAR 14] MAROTO J.M., MORAN M., "Detecting the presence of depensation in collapsed fisheries: the case of the Northern cod stock", *Ecological Economics*, vol. 97, pp. 101–109, 2014.

[MCF 73] MCFADDEN D., "Conditional logit analysis of qualitative choice behavior", in ZAREMBKA P. (ed.), *Frontiers in Econometrics*, Academic Press, New York, pp. 105–142, 1973.

[MES 08] MESNIL B., "Public-aided crises in the French fishing sector", *Ocean & Coastal Management*, vol. 51, no. 10, pp. 689–700, 2008.

[MER 10] MERINO G., BARANGE M., MULLON C., "Climate variability and change scenarios for a marine commodity: modeling small pelagic fish, fisheries and fishmeal in a globalized market", *Journal of Marine Systems*, vol. 81, nos. 1–2, pp. 196–205, 2010.

[MON 14] MONACO A., PROUZET P. (eds.), *Ecosystem Sustainability and Global Change*, ISTE, London and John Wiley & Sons, New York, 2014.

[MUL 11] MULLON C., FIELD J.G., THÉBAUD O., *et al.*, "Keeping the big fish: economic and ecological tradeoffs in size-based fisheries management", *Journal of Bioeconomics*, 2011.

[OST 90] OSTROM E., *Governing the Commons: The Evolution of Institutions for Collective Action*, Cambridge University Press, Cambridge, 1990.

[OST 10] OSTROM E., *La gouvernance des biens communs: Pour une nouvelle approche des ressources naturelles*, De Boeck, Brussels, 2010.

[PEC 01] PECH N., SAMBA A., DRAPEAU L., *et al.*, "Fitting a model of flexible multifleet–multispecies fisheries to Senegalese artisanal fishery data", *Aquatic Living Resources*, vol. 14, pp. 81–98, 2001.

[PEL 09] PELLETIER D., MAHÉVAS S., DROUINEAU H., *et al.*, "Evaluation of the bioeconomic sustainability of multi-species multi-fleet fisheries under a wide range of policy options using ISIS-FISH", *Ecological Modelling*, vol. 220, no. 7, pp. 1013–1033, 2009.

[PLA 85] PLATTEAU J.P., MURICKAN J., DELBAR E., *Technology Credit and Indebtedness in Marine Fishing: A Case Study of Three Fishing Villages in South Kerala*, Hindustan Publishing Corporation, Delhi, 1985.

[REV 91] REVÉRET J.P., *La pratique des pêches. Comment gérer une ressource renouvelable*, L'Harmattan, Paris, 1991.

[ROB 47] ROBBINS L., *Essai sur la nature et la signification de la science économique*, Librairie de Médicis, Paris, 1947.

[RUD 84] RUDELL K., AKIMICHI T. (eds.), "Marine Institutions in the Western Pacific", *Senri Ethnological Studies*, vol. 17, 1984.

[SCH 54] SCHAEFER M.B., "Some aspects of the dynamics of populations important to the management of commercial marine fisheries", *Bulletin Inter-American Tropical Tuna Commission*, vol. 12, no. 3, pp. 27–56, 1954.

[SCO 07] SCOTT A., "Phases in the evolution of property in sea fisheries", in BJORNDAL T., *et al.* (eds.), *Advances in Fisheries Economics*, Blackwell Publishing, Oxford, pp. 17–31, 2007.

[SMI 69] SMITH V.L., "On model on commercial fishing", *Journal of Political Economy*, vol. 77, pp. 181–198, 1969.

[SOL 86] SOLOW R.M., "On the intergenerational allocation of natural resources", *Scandinavian Journal of Economics*, vol. 88, no. 1, pp. 141–149, 1986.

[SUT 79] SUTINEN J.G., "Fishermen's remuneration systems and implications for fisheries development", *Scottish Journal of Political Economy*, vol. 26, no. 2, pp. 147–162, 1979.

[SUM 10] SUMAILA U.R., KHAN S.S., "A bottom-up re-estimation of global fisheries subsidies", *Journal of Bioeconomics*, vol. 12, no. 3, pp. 201–225, 2010.

[THI 12] THIAO D., CHABOUD C., SAMBA A., *et al.*, "Economic dimension of the collapse of the 'false cod' *Epinephelus aeneus* in a context of ineffective management of the small-scale fisheries in Senegal", *African Journal of Marine Science*, vol. 34, no. 3, pp. 305–311, 2012.

[THO 81] THOMPSON J.D., "Climate, upwelling and biological productivity, some primary relationships", in GLANTZ M.H., THOMPSON D. (eds.), *Resource Management and Environmental Uncertainty: Lessons from Coastal Upwelling Fisheries*, John Wiley, New York, pp. 13–33, 1981.

[VON 38] VON BERTALANFFY L., "A quantitative theory of organic growth", *Human Biology*, vol. 10, no. 2, pp. 181–213, 1938.

[WIL 80] WILSON J.A., "Adaptation to uncertainty and small numbers exchange: the New England fresh fish market", *Bell Journal of Economics*, vol. 11, no. 2, pp. 491–504, 1980.

[WOR 08] WORLD BANK, The sunken billions: the economic justification for fisheries reform, Washington, 2008.

4

Maritime Economy: Definition and Main Aspects

The maritime economy can be defined as the whole range of activities related to the sea, in other words whose conditions of development depend on the sea. This factual definition needs to be specified according to the degree of dependence: total, for the entire production mode, or partial, for the manufacturing of certain products or the supply of certain services. Furthermore, the definition raises a series of questions. A question of coverage: is it possible to list and delineate the whole range of "maritime" economic activities? A question of measuring the economic value of this range: is it possible to give a definition and a quantitative assessment of this range? By means of which indicators? A question of coherence: is the whole range composed of activities unrelated to each other? Or do they have commercial relationships and conditions of development (environmental, regulatory) leading, to a certain extent, to consider them jointly? The present chapter aims to address these issues on a European scale.

Capital for the development of the European Union (EU) and its coastal countries, driving force of the economic history of Europe where cities and port regions have played strategic roles (Genoa, Bruges, Ghent, Venice, London, The Hanse), the maritime economy is highly diversified: marine resources, international transport,

---

Chapter written by Régis KALAYDJIAN.

manufacturing industry, coastal and maritime services. The specificity of this economy refers, of course, to the marine environment, biological wealth, energy resources, chemical and physical properties and the space of displacement it offers, but also to its role as recipient of many pollutant emissions with terrestrial and marine origins. It also refers to the exiguity of coastal areas where the population density is particularly high[1]. In coastal and maritime areas, several management strategies may interact and have objectives which are barely compatible: industrial development, exploitation of recreational areas or the protection of the marine environment.

The maritime economy is not of negligible importance:

– maritime commerce provides 75% of the EUs external trade in volume and 50% in value[2] with connections to inland navigation and to land transport modes;

– coastal tourism is in steady growth and large coastal cities are in expansion;

– submarine cables are the backbone of international telecommunications;

– marine energy, mineral and biological resources constitute national targets that incite states to extend continental shelves.

These fundamental characteristics are found in all coastal countries and illustrate the impacts of maritime activities on the entire economy of a country or a region.

The current period is important to the maritime economy and policy of the EU and the member countries due to the implementation, a few years ago, of a set of initiatives related to the sea in industrial and environmental areas. The preparation of these measures required a number of economic assessments of industrial sectors and their

1 The coastal regions of the European Union represent more than 40% of the EU population. The "coastal regions" are the territorial units of the EU, roughly equivalent in size to that of French "departments", and of which at least half of the population lives less than 50 km from the sea. Source: Eurostat.
2 Source: Eurostat.

impacts on the marine environment. The present chapter focuses on this subject and aims to highlight:

– the main objectives of the European maritime policy;

– the knowledge of the economic impacts of maritime activities that the implementation of this policy requires;

– the economic assessment methods used, their benefits and their limits.

Before addressing the objectives of the European maritime policy, section 4.1 summarizes the main features of Europe's maritime economy. It is indeed essential to indicate in what economic context maritime policy is formulated and on what types of issues economic analysis is carried out during the preparation of decision making.

4.1. Overview of the European maritime economy

The maritime economy appears as a sample of the whole range of economic activities of a country or a region. It includes the primary sector (resource extraction), the manufacturing sector and the very diverse sector of services. These activities are carried out at sea or onshore, most of them requiring a double localization, at sea and onshore. The brief overview given in this section is limited to the main strictly marine activities: resources, transport and management of the marine environment.

4.1.1. *Marine resources*[3]

The marine resources exploited in the EU are energy, mineral and living resources. If the exploitation of energy and mineral resources is managed at the national level, commercial maritime fishing is the subject of a common fisheries policy (CFP).

The Treaty of Lisbon has introduced a legal basis specific to energy (article 194) in relation to key issues (efficiency, market, energy

3 Reference will also be made to vol. 3, *Ecosystem sustainability and global change*, in the "Seas and Oceans" set [MON 14a], Chapter 4, and the current book, Chapter 3.

supply, energy networks), but the exploited resources remain within the jurisdiction of states. Among these resources, offshore oil and gas are major ones for economic reasons. A total of 85% of crude oil extraction in the EU and 65% of natural gas are offshore. The activity is almost exclusively in Northern Europe, and marginally in the Black Sea. Exploitation in the Mediterranean, still low, is increasing significantly. In 2011, the offshore crude oil production of the EU exceeded 63 million tonnes (almost 11% of the total consumption) of which three-quarters was by the United Kingdom and a little less than 20% by Denmark. Gas production was over 90 Mtoe[4] the same year (a quarter of the gross consumption of gas of the EU), of which more than 50% by the United Kingdom and 20% by the Netherlands. But Norway was by far the largest European producer of offshore gas with 90 Mtoe, almost as much as the production of the EU[5].

The offshore oil and gas production of the EU was, in 2011, with a value in the order of 60 billion euros, roughly estimated from Brent[6] and European gas average spot prices. In volume, its tendency points downward, due to the largest producer, the United Kingdom. In Norway, the offshore oil production has dropped by 50% from 2001 to 2012, but that of gas has doubled over the same period.

Marine energies are at the research (e.g. ocean thermal energy conversion), pilot projects (wave energy) or the first commercial installation (current turbines) stage, apart from tidal energy and offshore wind farms:

– tidal energy has been present in France since the 1960s on the Rance estuary (with a 240 MW plant); only one project of this size has been carried out in Korea;

– offshore wind energy has reached an industrial stage, although with a higher unit cost than conventional thermal power production; its financial sustainability still depends on the buyback prices paid by

4 Toe: tonne of oil equivalent. Mtoe: million toes.

5 Source: Joint Research Center.

6 Brent: light crude from the North Sea. Its price on the spot market is a frequent reference price.

electricity transmission grids. Europe is the world pioneer and represented 90% of the installed power accumulated in 2012 with 5 GW (China 9%, Japan 1%) of which two-thirds are in the North Sea, 20% in the Atlantic and the rest in the Baltic Sea. The United Kingdom is by far the largest investor with 60% of the installed power, followed by Denmark (18%). The British effort will remain important as an additional 25 GW is the objective for 2020; the Asian effort will increase[7]. Without any offshore wind farms, France has just allowed 2,000 MW under construction off the coast of the Normandy, Brittany and Pays de la Loire regions. A new call for tenders for 1,000 MW is expected to be awarded in 2014. In Europe, the activity concerns bottom-mounted wind turbines[8] in shallow depth; floating turbines, adapted to the deep waters of Atlantic and Mediterranean, and more expensive, have not yet reached the industrial stage but are the subject of prototype development programs.

Deep mineral resources are not subject to a European management policy. Requests for the extension of the continental shelf of the different coastal countries beyond the 200 nautical miles – following the United Nations Convention on the Law of the Sea of 1982 – remain national matters. France, holder of a particularly important exclusive economic zone (EEZ) (world second largest after that of the United States with 11 million km^2), has a clear interest in acquiring rights to develop seabed resources in another zone of approximately 2 million km^2. The mineral resources, which these seabed zones potentially provide, are strategic: hydrothermal sulfides, cobalt crusts, polymetallic nodules, hydrogen, rare earths, biological and genetic resources, a detailed stock assessment remains to be performed (see [CON 13]).

Marine aggregates (sand, gravel and crushed stone) are mineral resources extracted near the coasts. Silica is predominantly used as construction materials for beach nourishment, the protection of coasts, construction fill and land reclamation and very marginally for market gardening. Limestone, in the minority, is used as a construction material and for soil amendment.

7 Source: *European Wind Energy Association*.

8 The tower-bearing structure is fixed on the ocean floor.

Marine aggregates are also managed at the national level and regulatory regimes may differ significantly from one country to another. They are mostly extracted at a depth of 20–30 m (up to 50 m in France) of the coastal areas of the North Sea and the Channel, and to a lesser extent of the Atlantic. Most of the coastal countries of Europe are producers with very different levels. According to estimates from the International Council for the Exploration of the Sea, the most important are the Netherlands (more than 60 million m^3 in 2011) and the United Kingdom (11 Mm^3: 20% of its total production of aggregates), these two countries are exporters. France[9] produces 4.5–5 Mm^3 (2% of its total production) and is importing. In comparison, the United States is under 10 Mm^3 and Japan[10] has been able to achieve approximately 40 Mm^3 in 2002. The Dutch level of production, around 20 Mm^3 before 2009, has exceeded 120 Mm^3 in 2009–2010 to decrease then, after the construction of a container terminal gained from the sea, today completed, in the port of Rotterdam.

The production of aggregates is faced with the growing difficulty to obtain authorizations for onshore extraction. Offshore extraction therefore offers an alternative with a variable importance depending on the country. Independently of the regimes of concession and licensing with a complexity variable from one country to another, offshore extraction raises issues related to its ecological footprint. The degradation of habitats and the effects of turbidity related to the process by hydraulic suction require to be studied to clarify intensity and persistence. The capacity of recovery of exploited areas seems to vary according to extraction process and location.

Finally, fish resources are the only marine resources managed at the EU level. The CFP (1983), inspired by the Common Agricultural Policy and revised every 10 years, is a policy implemented for the management of resources, but also for the development of the fishing sector. It has the objective for sustainable exploitation of fish resources and has four components: (1) the management of fishing quotas on the main commercial resources, granted to each member

9 Source: National Union of Aggregates Producers (UNPG).
10 Source: UNPG.

state; (2) the management of the production means by financing the adaptation of the European fishing fleet; (3) the organization of markets to stabilize the incomes of the producers; (4) the fishing agreements with third countries that, for the most part, consist of ensuring the access of European vessels to additional areas, against financial counterparties or third party access to European markets.

The EU is a large exploiter of seafood products with a little less than 5% of the world production (fisheries and aquaculture), but far behind China that represents a third of it[11]. The biggest European producers are Norway (more than three million tonnes), Iceland and Spain (more than a million). In the case of the EU, the activity is of economic significance: in terms of values of landed fish products, the production of the EU-28 exceeded 7 billion euros in 2011. Norway (more than 2 billion), Spain (1.7 billion), Iceland and Italy (more than a billion) accounted for the most important values. From 2007 to 2012, these values have decreased by 6% for the EU and increased by 4 and 20%, respectively, in the cases of Iceland and Norway. Therefore, two of the most important European fish producers, and of the most dynamic, are outside the EU and CFP.

The problem of the CFP is the non-achievement of its objectives: the decline of the commercial production of the EU for two decades (see [EUR 13b]) and the failure of the management of overcapacity have led the European Commission (EC) to propose a decentralization of the CFP [EUR 11c] through national multiannual management plans; the control of discards and the establishment of temporary and transferable fishing concessions within the EU are also proposed by the EC. More generally, on a global scale, the production of commercial fishery is slowly declining and has decreased below 80 million tonnes since 2008, not without effects on the commodity supply strategies of the seafood processing industry. Conversely, the human consumption of fishery and aquaculture products is recording a steady increase (more than 130 million tonnes in 2011), largely under the impetus of China's growth. The difference is made up mainly by the Asian freshwater aquaculture, but these figures suggest the risk of tension on the market of aquatic protein in the mid term and long term.

11 See also [MON 14a, Chapter 4].

4.1.2. *Maritime transport*

Nearly 80% of world trade by volume and 70% in value is carried out by maritime transport. This mode has accompanied a major evolution of the international trade of manufactured products due to containerization, which is generalized to most of the goods. From 1980 to 2013, the capacity of the world container fleet has increased 20-fold whereas that of the trade fleet has increased by about 140%[12].

The EU, through the member states, has acquired an important shipping fleet and active enterprises at the worldwide scale[13]. This should not hide the growing importance taken by the Far East (Japan, China, South Korea) in maritime transport and shipbuilding, but also by developing countries, now owners of a third of the world's fleet (Table 4.1).

Countries	Number of ships as of January 31, 2013	Capacity (dwt) as of January 31, 2013	Capacity (dwt) as of January 31, 2008
Greece	3,695	245	175
Japan	3,991	224	162
China	5,313	190	85
Germany	3,833	126	94
South Korea	1,576	75	38
Singapore	1,888	64	29
United States	1,943	58	40
Norway	1,908	46	47
Taiwan	814	44	26
Denmark	991	41	27

Units: number of ships (of 1,000 UMS[14] and more); million deadweight tons

Table 4.1. *Owned commercial fleet: the first 10 countries on January 31, 2013 (source: UNCTAD, Review of maritime transport, 2013)*

12 Source: United Nations Conference on Trade and Development (UNCTAD). Capacity here measured in "deadweight", which is a unit of weight; deadweight ton (dwt) being the unit of maximum load that a vessel can transport.

13 The first three world companies with regard to the capacity of the fleet in service, the transported tonnages and the turnover are European.

14 UMS (*Universal Measurement System*): new unit of transport capacity (gross tonnage or net) of vessels of more than 24 m engaged in international voyages.

The importance of Asia in the manufacturing sector has an evident impact on container-ship traffic on Asia-Europe and Asia-North America maritime routes but also, in a period of economic slowdown in Europe and in the United States, on intra-Asian and South-South routes, stimulated by the increase of the Asian consumption, and especially Chinese, of raw materials. Observers expect an increasing role of Africa in maritime traffic in the years to come, due to the economic growth of this continent; the EU remains for the moment the first trading partner of Africa before China.

The developments of trade in manufactured goods and in raw materials have a number of consequences on the growth of maritime ports. The first 10 containerized ports are Far East Asian ports – mainly Chinese – with the exception of Dubai (ninth). Rotterdam, first in Europe, is 11th worldwide. In Europe, the largest containerized ports are on the North Sea: Rotterdam (11.9 Mteu[15] in 2011), Hamburg (8.8 Mteu), Antwerp (8.6 Mteu) and Bremen (6.1 Mteu). On the Mediterranean, the main containerized ports have functions of transhipment relatively significant in their total traffic, beginning with Valencia (4.5 Mteu) and Algeciras (4 Mteu) but also, in the case of the Eastern region, Ambarli (3 Mteu) and Piraeus (2.7 Mteu)[16].

One of the objectives of the transport policy of the EC is to include the maritime transport in a strategy of intermodality toward a single European transport area: the EC seeks to promote Short Sea Shipping (SSS) at the scale of the EU (which represents, in tonnage, approximately 40% of the European freight[17]). One of the means consists of financial incentives ("Marco Polo" programs[18]) to implement "motorways of the sea" or door-to-door intermodal services, between two different member states, using SSS. These services are aiming to reduce traffic congestion, decreasing emissions

15 Twenty-foot equivalent unit (TEU): unit of container measurement. The standard container has a length of approximately 6 m (20 feet in American units). There are containers of different size (Mteu: million TEUs).

16 Source: *Journal de la marine marchande*.

17 Source: [COM 09a].

18 Marco Polo I (2003–2006, 102 Meuros), MP II (2007–2013, 450 Meuros) and MP III (2014–2020) will be launched. Regarding discussions about the effectiveness of these programs, see [COU 13].

of greenhouse gases and improving the accessibility of island or peripheral areas.

4.1.3. *Management of the marine environment*

The third domain mentioned here, the management of the marine environment, primarily concerns public action, but also involves the private sector. The subject takes a critical importance in the maritime policy of the EU as well as at the intercontinental scale because the maritime activities can have a worldwide range (exploitation of marine resources, transport and maritime services, sub-sea cables and oil and gas pipelines).

Records relating to maritime transport and offshore services, for example, assume that roles are shared between the EC and the International Maritime Organization (IMO) responsible for the intergovernmental cooperation on the maritime shipping management of technical aspects concerning safety, efficiency and the prevention and control of marine pollution by ships. These aspects are the subject of international conventions that the IMO updates in the light of technical developments to propose new ones if necessary; for example, the convention known as *marine pollution* ("Marpol") about the prevention of marine pollution by ships, *safety of life at sea* ("Solas") and *standards of training, certification and watchkeeping for seafarers* ("STCW"). The EC applies to the EU the IMO conventions of which it is a member as an intergovernmental organization having finalized a cooperation agreement. Three examples will be included that are particularly important to this distribution of roles between the EC and international agencies, in this case, the IMO: the prevention of oil spills, the reduction of air pollution and the standards of ship dismantling:

– the dossiers about polluting emissions from shipping, offshore platforms and maritime ports have both an international and a local dimension. They have given rise to a series of conventions of which the main one is Marpol. In the context of the latter, the funds for the compensation of damages due to oil pollution have been created (abbreviated: IOPC Funds) in 1971 and 1992, and introduced several regulatory measures about the safety of oil tankers: one of them concerned the withdrawal of single-hull tankers. This measure, like others, has been put in place and then tightened in response to serious accidents (Exxon Valdez, Erika, Prestige). In addition, the EC has responded to the Erika incident by the Erika I, II and III packages, aiming at the maritime safety of oil transportation, hardening the existing regulations (port state control, flag state responsibility, timetable for the withdrawal of single-hull oil tankers, traffic monitoring, investigations following accidents, inspection and classification standards) and creating the European Maritime Safety Agency;

– air pollutants in ship exhaust gas also fall under the action of the IMO: areas of controlled emission (*Emission Control Areas*) are put in place, particularly for the control of sulfur air pollutants (*Sulfur Emission Control Areas* (SECAs)), which will set limits on the sulfur content of fuel oil used onboard ships in these areas, starting from 2015. SECAs, defined in appendix VI of Marpol, also concern the EU: Baltic, North Sea and the English Channel;

– ship dismantling is driven by the profitability of shipping fleets, the prospects of freight markets, the prices of scrap steel and the current shift to the mandatory double hull for oil tankers. In 2012, the Indian subcontinent (Bangladesh, India, Pakistan) has processed 70% of the total capacity (in gross tonnage) sold for scrapping, of which 34% only to Bangladesh. The other important countries are China (second largest processing country with 22% of the capacity sold) and Turkey. The remaining share (5%) has been processed of these five countries[19]. The activity involves environmental and health risks that are increased in the open air facilities of South Asia; the old ships contain hazardous substances such as asbestos. At the level of the IMO, the convention of Hong Kong, adopted in 2009 and subject to ratification, foresees the control of the lifecycle of the vessel and an authorization for yards based on environmental criteria. For the moment, the Basel Convention (1989) is being applied, adopted under the aegis of the United Nations and aiming at the control of cross-border movements of hazardous wastes and their disposal. Pending the

19 Source: UNCTAD. Data concerning dismantled ships with more than 100 UMS.

entry into force of the Hong Kong convention, the problem of the EU in the short and medium terms is to manage a significant and growing demand for scrapping flag or owned vessels of EU countries, given that its own dismantling capacity would not suffice[20]. The EC seems to have few alternatives to propose in terms of recycling capacity with sufficient environmental standards, apart from the promotion of "clean dismantling" for state ships and commerce ships [EUR 08].

In contrast, an environmental matter falling under the jurisdiction of the EC, to which the rest of the chapter will frequently refer, is that of the quality of coastal and marine waters, which involves the use of fresh, coastal and marine waters, onshore and in EEZs of member states. The legislation in force foresees a return to the "good environmental status" of fresh and coastal waters in 2015, and 2020 for marine waters.

4.1.4. *The diversity of the maritime economy*

This brief overview has prioritized the activities at sea. It would have been able to extend to other, not less important sectors, whose markets are linked to those which were just mentioned: the manufacturing sector (ship and boat building and repair), financial services (insurance and bank), tourism (cruises, accommodation and marinas) and major public services such as defense and marine research.

As limited as it might be, this sample of activities illustrates the diversity of maritime activities' development process and modes of regulation. It shows in what economic and regulatory context the European authorities seek to formulate a maritime policy.

This policy requires an analysis of the maritime economy on homogeneous bases, as several elaborated studies have tried using simple economic indicators, applied to all the maritime activities: this

20 In 2007, the EC forecasted an insufficient capacity of demolition in the EU and in the OECD countries [COM 07a]. The underutilization of the ship recycling capacity in the EU, Turkey and China, with higher environmental standards, is also due to the conditions offered by the scrapping yards of South Asia (steel redemption price).

issue is discussed in the following section. These studies give much information on the analysis of the maritime economy, its interest and difficulties.

4.2. The European maritime policy and its economic information requirements

The European maritime policy includes a series of initiatives from the EC regarding the management of maritime activities, but without a specific legal basis in the treaty of the EU. Globally, this maritime policy has been aligned with the objectives of competitiveness and sustainability, formulated by the EU in a very broad framework:

– the Lisbon strategy (2000) is based on objectives of competitiveness and employment growth by 2010, in response to the globalization and the aging of European populations[21];

– the Gothenburg strategy [EUR 02] introduces objectives of sustainable growth, job creation and social cohesion.

The important stages of the maritime policy include:

– strategies that reach far beyond the maritime policy but that contribute directly to it: this is the case of the strategy of the trans-European transport networks (TEN-T), which involves notably the modal transfer of road systems to maritime/inland waterway; the motorways of the sea mentioned above are now included in the TEN-T;

– initiatives specifically maritime such as:

- the CFP referred to in the previous section and put in place in 1983;

- a multisector strategy: the integrated maritime policy (IMP) that, according to the blue book of 2007[22], reiterates several previous initiatives, including the CFP and the transport networks, and proposes an overall consistency;

21 See conclusions of the Lisbon European Council Presidency, March 23-24, 2000. Evaluation of the Lisbon strategy: [SEC 10].
22 See [COM 07b]. 2008–2010 action plans [SEC 07].

- an environmental strategy: the "marine strategy" framework directive (MSFD)[23], adopted in 2008, which has as its objective the achievement of a "good environmental status" of marine waters by 2020.

The multisector policy orientations, presented in the framework of the "blue growth", are part of the 2020 strategy of the EU and identify the sectors with high growth potential as a contribution to competitiveness and job creation.

These various political initiatives are calling for economic evaluations: evaluation of policy measures (before and after their implementation) and the economic sectors to which these measures apply. These evaluations, requested by the EC, have two substantive reasons: on the one hand, they inform the EC on facts and economic indicators on the sectors where data may be missing, and on the other hand, they are a key instrument of negotiation with the member states.

This is the case of the CFP that needs regularly updated information about the costs, incomes and employment of fisheries. A framework for the collection of fishing data was the subject of a series of European regulations[24]. The recitals of the 2008 Regulation remind of the usefulness of having "data explaining price formation" and "which may facilitate an assessment of the economic situation of fishing enterprises, aquaculture and the processing industry, and employment trends in these sectors". Article 9 of the Regulation specifies that the data should permit "impact assessments of measures undertaken, or proposed to be carried out".

This is also the case of the European strategy of the TEN-T. In the wake of the conclusions of the European Council of Gothenburg, the 2001 white paper on transport [EUR 01] emphasizes mainly on the objectives of European integration with regard to trans-European networks, road congestion and environmental impact management, safety, linking in outlying areas and enclaves and accompanying

23 2008/56/EC of the June 17, 2008, JOCE L164/19, June 25, 2008.

24 Regulation (EC) 199/2008 of the Council of February 25, 2008. Regulation (EC) 665/2008 of the EC (July 14, 2008).

measures to minimize the social impacts of the economic growth. The revision of the transport policy of 2010 [EUR 11a, EUR 11d, SEC 11] is more focused on the issues of energy and economic efficiency. It aims at the modernization of infrastructures, propulsion systems and fuels; the intermodality of logistics chains; the elimination of barriers to market access and the distortions of competition in what it projects as a "Single European Transport Area".

Economic evaluation has been an essential tool: the expert reports made around 2006–2007 for the EC consisted of a prior evaluation of TEN-T multiannual programs[25]. These numerous reports have focused on the economic impacts of transport investments on the competitiveness of the users, the efficiency of the existing transport infrastructures themselves, employment and distributed income. As these reports have pointed out, the analysis of the effects of the TEN-T strategy would require a detailed economic database on the impacted economic sectors as well as models of impact assessment.

The same observation can be made for the IMP in 2007. In its principles, it endorses the CFP as well as the maritime part of the TEN-T strategy, but goes beyond. It aims at a framework comprising management objectives for the marine environment, coastal zones and marine fisheries; the promotion of industrial innovation, maritime clusters and marine sciences; and the development of European transport networks without barriers in the areas of maritime transport and seaport logistics.

To address this broad spectrum, the IMP Blue Book (see above) emphasizes on the need for basic economic information on maritime activities. Action 6.5 of the 2008–2010 Action Plan aimed at the construction of a database "for maritime areas and coastal regions" as part of "follow-up actions". The tendered specifications for the construction of this database [EUR 07c] emphasized the operational orientation of the tool requested by the EC. More detailed operational data on maritime activities and their impacts on local resources should allow informing the analysis of marine and coastal uses and the

25 See [ECO 07, FRA 06a], annex 7 [FRA 06b].

impact studies necessary for the preparation of the measures of maritime policy.

Shortly after the IMP, the MSFD (see above) was adopted. "The environmental pillar" of the IMP according to the term employed by the EC, it follows the water framework directive (WFD)[26]. Its main objective is the "good environmental status" of marine waters, just as the WFD was aiming at the good environmental status of fresh, estuarine and coastal waters[27]. Coherence was needed between these two texts, given the adjacency of marine waters and coastal waters, and the criteria of good condition have been specified in the MSFD. Coherence is also required with the regulatory texts aimed at the good condition of biological and environmental resources.

The MSFD requires economic evaluations as well. The return to the "good condition" raises the problem of costs: curative costs of restoration and preventive costs relating to the behavior of upstream uses, and it is at this level that the contribution of the economic analysis is required. Article 8.1 of the directive thus provides an economic and social analysis of "the use of waters" of marine regions and subregions as well as the "cost of the degradation of the marine environment" (article 8.1c). It concerns the initial assessment, before the implementation of the directive. Article 8 is complemented by article 17, which defines the update of marine strategies and the initial assessment every 6 years.

These works of economic evaluation have an operational objective. The programs of marine strategy of the member states developed for the implementation of the MSFD will have to take account of certain cases. These cases would be those where the costs of the measures aimed at the achievement of the "good condition" would be "disproportionate taking account of the risks to the marine environment, and provided that there is no further deterioration" (MSFD, article 14.4): these cases may be considered as derogatory.

26 2000/60/EC of 23 October, 2000. OJ L 327/1, 22 December, 2000.

27 "Coastal waters: the surface waters on the landward side of a line in which every point is at a distance of one marine mile on the seaward side from the nearest point of the baseline from which the breadth of territorial waters is measured, extending where appropriate up to the outer limit of transitional waters" (articles 2–7).

After all, an economic and social assessment would compare for the development of management measures:

– the economic value produced by each of the uses;

– the cost of degradation avoided by the achievement of the good condition;

– the cost of the return to good environmental status due to the constraints imposed on water uses and the regulator.

Sections 4.7 and 4.8 discuss the methodological aspects of the initial assessment.

As the most recent step including a dimension of maritime policy, the "Europe 2020" initiative [EUR 10] combined the objectives of the Lisbon and Gothenburg strategies that are competitiveness and competition associated with the protection of the environment. The "Blue Growth" strategy [EUR 12] is part of this initiative: it is the European strategy for the sustainable development of the blue economy, this latter being defined as the set of activities related to the sea and corresponding to the maritime economy considered here. The priority sectors are selected in relation to their potential for innovation and their potential contribution to European growth: renewable marine energies, aquaculture, tourism and coastal cruises, marine mineral resources and marine biotechnology inter alia. In terms of economic evaluation, the preparation of the "Blue Growth" strategy has relied on [ECO 12], a study of the various maritime activities, their context and their growth potential.

The policy initiatives, summarized in this section, illustrate the key role of economic analysis to support EC decisions. Depending on projects and initiatives, the approach of the EC is oriented toward two types of issues:

– the issue of the impacts of maritime activities on employment and the production of their suppliers and users: the CFP, the IMP and the MSFD require specific data collection and key economic assessments;

– the issue of social costs induced by coastal environment degradation and the overexploitation of marine resources.

These two issues are essential and raise that of the choice of adequate evaluation methods. They constitute the major components of the approach developed in this chapter and are discussed in the following sections.

4.3. Sector-based approach to the maritime economy

The analysis of the economic impact of maritime activities is discussed here. It requires, first and foremost, a reliable set of data on their economic significance, notably a number of fundamental indicators: turnover, value added, employment, distributed income. Such an objective raises the question of building a "maritime satellite account". This would consist of structuring the information about maritime activities by means of a set of indicators periodically updated consistent with the national accounts.

Satellite accounts vary from one country to the other in numbers and in sectors concerned (for example in France: transport, environment, tourism, education, health, social protection). Some of them are the subject of international recommendations on form and content (for example tourism satellite account: [UN 08]). In the United States, it was in the 1970s that a model of maritime satellite account[28] was proposed; a number of rather more applied exercises have then been made in several countries, starting with the United States.

The first evaluations of the maritime economy have been launched in the 1970s in the United States, with the studies of the *Bureau of Economic Analysis (BEA)* on the contribution of maritime activities to

28 See [PON 80]. For an accounting framework extended to coastal activities, see [LUG 91].

the U.S. GDP[29]. The works of this federal agency in charge of national accounts have given rise, in the same period, to the construction of the *Economics: National Ocean Watch* (ENOW) database managed by the *National Oceanic and Atmospheric Administration* (NOAA) and targeted on the accounts of maritime activities. ENOW uses the accounting data of the BEA and the employment data from the *Bureau of Labor Statistics*. Since the end of the 1990s, this database has allowed the emergence of academic works funded by the NOAA and coordinated within the *National Ocean Economic Program* (NOEP) from the University of Southern California[30], notably economic reports about maritime activities as well as the online database of the NOEP, updated annually. The NOAA also has published a summary [BOO 12] on the maritime economy of the United States, assessing the six major activities (living resources, offshore mineral resources, shipbuilding, transport, tourism, maritime works) by means of standard indicators (number of businesses, value added, employment, distributed income). NOEP's and NOAA's database provide not only data at the federal level, but also estimates of accounting indicators by state, activity and for coastal counties[31].

In the 2000s, the Canadian federal government and the provinces have commissioned commercial consultancies to carry out several studies on the maritime economy of the country [GAR 09a] and coastal provinces [ADE 12, GAR 09b]. These studies cover either the whole of the maritime economy or a selection of sectors, using accounting indicators: turnover, value added and employment. The contributions of Australia [ALL 05] and New Zealand [STA 06] were also published in the 2000s.

Some works of the same type have been launched in the 1990s in Europe, first at the level of states, and then at the level of the EC. Among the numerous studies, the main ones are as follows:

29 GDP: gross domestic product, sum of the gross value added of the sectors of activity of an economy. The GDP is defined at the national level and for the sectors of activity filled by national statistics.
30 About history, see [KIL 00]. For the latest report of the NOEP on the U.S. maritime economy, see [KIL 09].
31 The NOEP makes econometric estimates. Counties, administrative subdivisions of states, are of varying sizes, more or less comparable to those of French "departments".

– the first works published by the United Kingdom and Italy dating back to the 1990s: the global analyses of the British maritime economy [PUG 08] were followed by more specific studies on the services related to port and maritime transport [OXF 13]. In Italy, the reports published periodically since 1998, such as [CEN 11], focus on a broad array of activities;

– in France, the French research institute for the exploitation of the sea (Ifremer) has published a first assessment of the maritime economy in 1994, which has been subsequently followed, since 1997, by the periodic reports "French marine economic data" ("*Données économiques maritimes françaises*"): [KAL 10]. The Sea and Coastline National Observatory, established in 2012, also uses data from public statistical services and "information systems" such as the water information system to provide a set of environmental, economic and population indicators, responding in part to regulatory requests such as those of the MSFD;

– in Germany, reports have been published on Schleswig-Holstein in 2005 and 2008 [JAR 08], the region where maritime activities contributed with more than 12% to the regional value added in 2006, while the national averages are at around 1.5%;

– with regard to the maritime industries of the Netherlands and Norway, a study [WIJ 03] was issued in 2003 on the initiative of industry associations in the two countries. In 2010–2011, studies were published on Ireland [MOR 10] and Spain [INN 11];

– at the European level, similar projects date from the same period. Through the initiative of the EC, the first study [POL 00] was carried out at the end of 1990 on Europe's maritime economy in 1997, assessed by the activity and country of the EU-15 and Norway, according to turnover, value added and employment. A second study [IFR 09] has been carried out in the framework of the 2008–2010 Action Plan of the IMP (see above). This work consisted of a general review of the available (and missing) data from the national statistical institutes and complementary sources to inform a very large set of maritime-related activities and accounting indicators.

As shown by the above-mentioned studies, there are at least three types of publications that differ in their objectives: the studies of private consultants carried out at the request of public authorities in support of maritime policies or for information purposes on the maritime economy; the studies of universities and public agencies for research, the purpose of producing knowledge and reuse the results in research projects or supporting public policy projects; the studies of private consultants commissioned by firms seeking to publicize their activities and their markets.

From a methodological point of view, these publications, of which the number remains modest, are divided into two types:

– studies that are based on a broad concept of maritime economy (the United States, UK, Italy, France) specifically comprising marine activities (see section 4.2) but also manufacturing, construction and a large number of coastal services, including tourism, accommodation, catering and local transport;

– studies that rely on a restricted coverage of the maritime economy limited to "strictly marine" activities (e.g. fisheries and shipping) and to goods and services related to maritime transport (Netherlands, Norway, Australia, New Zealand).

They use official statistics from the national statistical institutes and, for the most part, complementary sources: statistics of companies and industry associations and consultancy databases. The frequent problem of complementary sources relates to their data collection methods, poorly known to the end user: the resulting data can present bias difficult to assess (information from members of industry associations or commercially sensitive databases).

4.4. Maritime economy coverage

A good way to discuss the coverage of the maritime economy is to characterize the different categories of activities in terms of the links they have with the marine environment, coastal regions and the other categories of activities. The following categories can thus be distinguished:

– the “specific” activities that depend on the sea. They exploit:

- living marine resources, minerals and offshore hydrocarbons,

- the physical and chemical properties of marine waters (maritime transport, ports and port services to vessels, power plants using the sea as a cold source, renewable energies, defense, operational oceanography[32], marine route surveys, coastal water monitoring),

- the coastal and marine spaces (dredging and construction of port infrastructures, laying and maintenance of cables, oil and gas submarine pipelines, other maritime works),

- coastal landscapes and seascapes (marinas: exploitation, monitoring and berthing; local activities of maritime transport such as coastal tourism shuttles),

– the activities “linked” to the maritime economy are in close relation to the above specific activities in terms of production process: they are suppliers or clients, but not necessarily located in coastal areas. They are subdivided into “essential” activities and in “complementary” activities:

- the “essential” activities do not exploit directly the marine environment, but are key for the markets of specific activities and vice versa. They include the downstream services of the seafood chain (first sale/auction markets, wholesale fish trade, distribution and processing), ship and boat building, repair, maintenance and dismantling, the manufacturing of civil and military naval equipment, the manufacturing of pleasure boat equipment, nautical sports and recreation, offshore oil and gas support services, freight port (handling, warehouses, logistics, freight forwarders, agents and consignees) and passengers services and maritime services (sale and rental of commercial vessels, ship supply, brokering, classification, insurance); accommodation and catering; services related to boating

32 Operational oceanography: collection and provision of data in real time about the environment and the state of marine waters (see *Development of Marine Resources* in the “Seas and Oceans” set [MON 14b]).

(including rental and sale of boats and equipment, clubs and access rights to water sports) and miscellaneous expenses of stay on coastal tourist places, including local transportation and cruise tourism with stopovers; signaling and maritime safety, training, protection of the marine environment (collection of marine debris and hazardous waste at sea);

– "complementaries" are clients or suppliers of specific activities, each of these two categories providing markets to another category: biotechnology, textiles and professionals clothing; various naval equipment, whose prospects are maritime and non-maritime, such as electrical systems, coatings and paints; river work and navigation, travel agencies and other tourism services (guides, ticket offices, etc.), wastewater and urban wastes treatment. This is also the case of a number of more specialized activities (for example legal services, shipping press and professional associations);

– several coastal activities are without any relation with the maritime economy itself, but can be implemented in port, urban or coastal rural areas: coastal agriculture, building and civil engineering industry, wholesale and retail distribution, people services (for example health, banking, legal and trade), manufacturing in port areas (petrochemicals, steel, refinery, gas, construction materials), specialized logistics.

This description combines criteria of industrial and geographical types characterizing the production of goods and services and the place of production. Table 4.6 in section 4.10 describes the coverage of maritime economy by the national reports mentioned in section 4.3. It was noted earlier that some studies are more selective than others in their coverage. But this variable selectivity also exists in the coverage of certain activities. This is the case of coastal tourism that includes, on the one hand, nautical activities including boating and all related expenditure, and, on the other hand, stay expenses including accommodation, catering, local transportation and subsistence expenses. If the majority of studies seek to cover yachting and boating, activities classified here as "specific", they are significantly less numerous to take into account stay expenses (the United States, Italy, Great Britain, Canada, France).

4.5. Maritime economy: sector-based approach and methodology issues

In the studies taken into account in section 4.3, each maritime activity is assessed using a set of economic indicators. In this regard, the most systematic and transparent coverage in terms of sector-based statistics is that of national accounts. The sectors are identified and hierarchized according to a classification of activities developed at the national level and harmonized at the international level[33] by the United Nations Statistical Division: the classifications in force in the different countries follow that of the United Nations while taking into account the characteristics of the national industry. Therefore, the French classification of economic activities (NAF) is the application to France in an adapted form of the Statistical Classification of Economic Activities in the European Community (NACE), periodically updated; at its finest level, the NAF includes 732 activities.

The recourse to this classification is facing a difficult challenge: the activities considered here as maritime are not all specifically identified in the European and national classifications. If some of these activities are defined in the NAF and directly informed by a set of economic indicators (accounting statistics, employment and foreign trade), others are a much more numerous part of broader categories of the NAF, which include maritime or non-maritime activities:

– the first group includes, for example, fishing, shipbuilding and maritime transport;

– the second group includes the extraction of marine aggregates, all types of naval equipment, the manufacture, laying and maintenance of submarine cables, the offshore oil and gas industry and numerous port, tourism, legal, engineering and consultancy services;

– some activities can only be regarded as maritime if their production units are close to the shore or harbor areas, such as hotels

33 *International Standard Industrial Classification* (ISIC), periodically updated by the UNSD.

or port logistics (if it is agreed to include them in the maritime economy); such a restriction assumes, in order to inform them, to resort to local statistics.

So, an important share of the maritime economy is difficult to apprehend by the usual national account indicators. This results in two types of trade-off: the coverage versus accuracy trade-off and the resolution trade-off.

– *Coverage versus accuracy trade-off*: the maritime activities merged into broader sectors of national accounts and in the NAF are obviously difficult to assess using standard economic indicators. At best, these can be estimated. Regardless of the reliability of estimation methods, a margin of substantial error is predictable, given the lack of means to appreciate it. Therefore, a wide coverage of the maritime economy is only possible at the expense of accuracy. Limiting the scope to accurate and reliable accounting data presupposes excluding a significant part of the maritime economy.

The exclusion of any estimated indicator is the option chosen by the maritime study of the New Zealand institute of statistics [STA 06], which takes account only of the maritime activities informed by accounting indicators. This is the reason for the selective coverage. Eurostat, the EC Directorate General in charge of the elaboration of community statistics, retains the same option in its analysis of the evolution of maritime activities over a period of about 10 years [COL 13].

– *Resolution trade-off*: the coverage of the maritime economy endorses a geographical dimension because its components are, for the most part, confined to coastal areas. However, the accounting indicators characterizing the economic activities are, most often, informed at national level only (turnover, value added, etc.); at a regional and local level, they lose their meaning. Eurostat notably provides data relative to regions of member countries, but only for large groups of activities, not for the finest subdivisions of the NACE. There is consequently a trade-off between sector and spatial resolutions: the geographic accuracy of the accounting data can only be obtained at the price of a lesser sector resolution. This trade-off is illustrated by the maritime database of NOEP where the indicators relating to the counties[34] are only given for groups of maritime activities, with an accuracy level markedly lower than for states; these are estimates, as reported above.

Some exceptions confirm the rule: in France, the INSEE national statistical institute develops an annual local database, observed and not estimated, of employment at the place of work, remuneration and numbers of establishments by statistical activity and municipalities[35]. Despite this tool, the resolution conflict remains because statistical confidentiality applies when, for a given sector of the NAF, the territorial units contain fewer than three companies or establishments or when a single company or a single establishment contributes to more than 85% of the turnover; lifting statistical confidentiality requires using a wider sector or a greater geographical unit at the cost of accuracy.

The two trade-offs reported above demonstrate the absence of a standard delimitation of the maritime economy. The choice of coverage assumes a conventional aspect; it depends on the economic field that is required to inform or that is estimated potentially concerned by political initiatives. It also presents a practical aspect: an activity will be taken into account depending on the existence and the quality of the data. A maritime economy database is, in the end, a compromise between a selection judged to be consistent of activities and a level deemed eligible for data quality.

4.6. Sector-based approach to the French maritime economy

In 2011, the assessment of the French maritime economy has been put on line relatively to years 2009 and earlier[36]. The coverage taken

34 County: administrative subdivision of the American federated states, of a variable size, roughly comparable to that of a French “department”.
35 “Connaissance locale de l’appareil productif” (CLAP) – (local knowledge of the productive system) database.
36 www.ifremer.fr/demf. The 2013 update referring to 2011 was not available during the drafting of this chapter.

into account[37], enlarged compared to the first issue of the "French marine economic data" report, is now aligned as per the database performed for the IMP (see section 4.3). The main features of the study are:

– characterizing, by means of accounting statistics of INSEE (turnover, gross value added at basic prices, and number of employees or FTEs, the maritime activities fully identified by a NAF code;

– estimating the accounting indicators of maritime activities included in wider NAF sectors by means of indicators and ratios from INSEE and industry-sourced estimates; these estimates are to be interpreted with caution;

– characterizing maritime activities with complementary, non-monetary indicators, for example tonnages of transported cargo, tonnage of port throughput and overnight stays in hotels;

– characterizing public services by a set of specific indicators, including the budgets allocated to these activities but also personnel expenses; these are used as an estimate of the value added in the absence of information on capital amortization. Specific indicators may also include, where appropriate, non-accounting complementary indicators, for example the number of students in training programs in a particular discipline;

– the study thus proposes a general survey of maritime activities, their short-term development and their place in the international competition. It performs an overall estimate of the number of jobs and maritime value added.

Table 4.2 gives the main results of the assessment for 2009. These indicate a value added of 22.6 billion euros for approximately 460,000 jobs. These figures correspond to approximately 1.5% of the French employment and GDP. This proportion is equivalent to what is observed in other countries, for example the United States. There are countries with very long stretches of coastline, where maritime activities occupy a relatively more significant place; this is the case

37 See Appendix 1.1 for the sectors taken into account.

for Norway (7%), the United Kingdom (3–4%), the Netherlands (3%) or Italy (1.5–2.5%)[38].

	Value added (2009, %)	Employment (2009, %)
Tourism	46.3	46.8
Shipbuilding	7.5	9.7
Seafood products	10.9	9.9
Maritime transport	6.1	13.0
Public sector	14.3	11.9
Offshore oil and gas industry	10.9	5.4
Other	3.9	3.2

Table 4.2. *French maritime economy in 2009 (sources: INSEE, state administrations, industry associations and businesses; Ifremer estimates from INSEE data)*

These figures highlight the very large share of coastal tourism in the French maritime economy. This activity represents almost half of the whole. Four other groups of roughly comparable significance are as follows: civil and military shipbuilding (including boat building, ship and boat repair, naval equipment), seafood products (fishing, aquaculture, trade, processing), maritime transport (including maritime ports and inland navigation) and public service (national navy, state intervention at sea, education, research). In 2009, maritime transport and shipbuilding have been particularly affected by the economic crisis. Their contribution to the total value added has been well below than what it was in previous years, contrary to their contribution to employment in 2009, which has not significantly declined. Evaluated from data from the French Petroleum Institute, the offshore oil and gas service industry equally constitutes an important activity in which labor productivity appears with high values. The public sector, for its part, is strongly influenced by the evolution of the number of personnel in the national navy, which constitutes its major part; it has a growing relative importance during periods of crisis (this

38 These figures are extracted from the reports cited above, with specific perimeters in the case of the maritime economy.

was the case in 2009) and decreasing in favorable periods during which its production share is around 10%.

One of the aspects of the French maritime economy is the high sensitivity of several activities to external markets due to the modest size of their domestic market: cruise ship building, marine insurance and offshore oil and gas services, for example, are large exporters. This sensitivity is illustrated by the significant variability in export rates as a function of the periods of growth, slowdown and recession.

Data in Table 4.3 have to be compared to the average export rates of the French industry (Source: Banque de France): 17.8% and 18.5%, respectively, for 2009 and 2010; approximately 34% over these two years for the manufacturing sector only; 31.5% for transport.

Activities	NAF code	Export rates 2010 (%)	Previous years export rates
Fish processing	10.20Z	8.20	From 13 to 16% in 2001–2009
Shipbuilding	30.11Z	45.40	From 30 to 45% in 2006–2010
Boat building	30.12Z	46.50	From 50 to 65% in 2001–2009
Ship repair	33.15Z	0.80	From 35 to 50% in 2001–2007, close to 0 after 2007
Passengers maritime transport	50.10Z	0.04	Total freight and passengers: from 60 to 80% in 2001–2007, 22% in 2006, close to 0 after 2007
Freight maritime transport	50.20Z	0.55	

Table 4.3. *Export rate of French maritime activities (source: INSEE)*

4.7. From a sector-based approach to a maritime basins approach

A maritime basins approach, whose benefits have been discussed in [IFR 09], tends to take its shape from the most recent studies. In principle, it consists of assessing the economic activities dependent on a common maritime basin (for example marine subregion mentioned previously, or an area with smaller dimensions) from the perspective of the exploitation of resources (biological, energy, mineral, seascape), maritime transport, coastal tourism, cruise ports or shipyards, among others. It is also about assessing the exchange of goods and the movement of people between coastal regions of the basin. The aim is to analyze how such a basin works and which co-operation and competition relationships exist between coastal regions and between employment zones. Such an approach leads to performing a number of regional comparisons regarding the weight of the different economic sectors, transport and networks infrastructure (cables, oil and gas submarine pipeline), demographic and social characteristics (population density, income levels, occupation categories, active and non-active population). The difficulty of such an approach is that it requires economic data about the various economic activities, geographical data about the delimitation of maritime regions, marine data (biological resources, mineral and energy, marine and terrestrial transport infrastructures) and data from interregional trade.

Given the significant amount of data needed and their level of accuracy, the feasibility of the exercise is not assured. It is nevertheless necessary to cite two examples of application, which are interesting because of their limits. They have allowed exploring what economic data are available for the study of maritime basins.

4.7.1. *Economic and social assessment of the MSFD: use of marine waters*

The initial assessment of the MSFD conducted in 2011 (see section 4.2) aimed to compare the value produced by the use of marine waters, the cost of environmental degradation and the cost of the return to good condition.

Article 8, already quoted, refers to the initial evaluation. Article 13 describes the program of measures that member states must define and provide that they "shall ensure that measures are cost-effective and technically feasible, and carry out impact assessments, including cost-benefit analyses, prior to the introduction of any new measure".

The cost of the impacts on the environment is therefore compared to that of the proposed management measures, aiming at the "efficiency" of the measures. Derogations are referred to in article 14, which stipulates that the member states "shall not be required ... to take specific steps ... when the costs would be disproportionate taking account of the risks to the marine environment, and provided that there is no further deterioration". The concept of "disproportionate cost", also present in the WFD, allows granting to the member states a latitude of appreciation of their priorities.

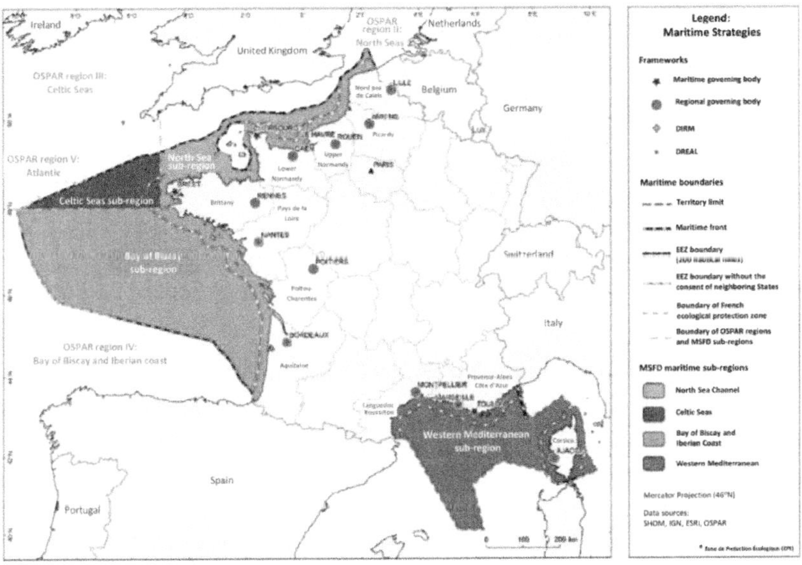

Figure 4.1. *Delimitation of submarine regions concerning France's seafronts (source: Agency of marine protected areas, Ifremer) (see color section)*

In the absence of an assessment method imposed by the directive and the associated guidelines, a working group on economic and social assessment (WG-ESA) made up of experts and representatives of end users has published a non-constraining guidance document

[WGE 10], which offers several work directions. One of them, the marine water accounts approach, consists of a sector-based analysis of water uses. The approach is close to the national reports outlined in the preceding sections, but obliged to limit the assessment of indicators to each submarine region separately[39]; it is comparable to a sea-basin approach.

In the case of the initial assessment for France, these subregions are the English Channel–North Sea, the Celtic Seas, the Bay of Biscay and Western Mediterranean (see Figure 4.1 above). The social and economic assessment (November 2011–June 2012) has given rise to the preparation of files, by maritime activity and marine subregion, characterizing the activity through a set of accounting indicators at the national level (key figures), a description of its situation in the submarine region and its regulations and impacts on the marine environment (Table 4.4). Several other member states have also adopted an accounting approach of maritime sectors[40].

The files have been attached to the "action plan for the marine environment" (APME) of each marine subregion, in accordance with the MSFD. The three plans, supported by the environment ministry's local services of the marine subregions (English Channel–North Sea, Mediterranean, whole Bay of BiscayCeltic Seas), are the subject of an online documentation, including a detailed description of the APME and a initial assessment document, which brings together the revised sector-related files[41]. The maritime basin sector approach is thus applied to an operational case; in principle, the work should allow comparing the value produced by the activities, the costs of environmental degradation and the costs of a strategy of return to the good status of marine waters, coordinated at the level of the marine subregion as far as it is technically required.

39 See article 4 of the MSFD concerning marine regions and subregions.

40 Belgium, France, Germany, the Netherlands, Ireland, Norway, Spain and UK according to an intermediary note from EFTEC (2012).

41 See [MED 12a] and [MED 12b] for the initial Mediterranean and English Channel–North Sea assessment. The APME for the Bay of Biscay and Celtic Seas was in preparation during the drafting of the present contribution.

		English Channel–North Sea	Celtic Seas	Bay of Biscay	West Mediterranean
	Use of waters				
1	Subsea cable manufacturing, laying and maintenance	x	x	x	x
2	Merchant and naval ship building, repair and dismantling	x		x	x
3	Defense, as limited to activities covered by the MSFD	x	x	x	x
4	Marine aggregate extraction	x		x	x
5	Initial and continuing education	x		x	x
6	State activities, state intervention at sea	x	x	x	x
7	Tourism and leisure				
7.1	Beach-related recreation	x		x	x
7.2	Boating and nautical sports	x		x	x
7.3	Recreational fishing	x		x	x
8	Offshore oil and gas services	x	x	x	x
9	Coastal power plants and marine energy	x		x	x
10	Coastal and marine environment protection	x	x	x	x
11	Research and development	x		x	x
12	Seafood sector				
12.1	Mariculture	x		x	x
12.2	Commercial fishing	x	x	x	x

12.3	Seafood processing and trade	x		x	x
13	Maritime financial services	x		x	x
14	Maritime and river transport	x	x	x	x
15	Maritime work and civil engineering	x		x	x
16	Other coastal activities, and activities with significant impacts on marine environment				
16.1	Agriculture	x		x	x
16.2	Manufacturing and art and craft activities (not elsewhere included)	x		x	x
16.3	Coastal tourism	x		x	x
16.4	Loss of natural landscape and increasing artificialness of coastal zones	x		x	x

Table 4.4. *Economic and* social analysis *France (MSFD): use of waters (source: Agency of marine protected areas, Ifremer)*

The assessment of marine subregions has come across a number of difficulties known by the national reports (see section 4.5) because of the required local character of the information:

– resolution trade-off: the difficulty in accompanying the description of the situation by marine subregion by local indicators has been partially compensated by the local data of INSEE; non-monetary indicators have also been used: for example, tonnages generated, landed, transported and tourist overnights;

– in the case of maritime activities included in broader sectors, such as seaside tourism and water sports, education and marine research, sector data were most often limited to non-monetary indicators.

4.7.2. *The Atlantic maritime basin: a European project*

One example of a maritime basins approach is given by the research project, Marnet (2012–2015). This project, funded by the Atlantic Area component of the Interreg program, itself funded by the European Regional Development Fund[42], has been launched with the aim of gathering information and economic data useful to the preparation of the decisions relating to maritime and coastal policy. Inspired on the national reports on maritime activities, it consists of implementing an economic, demographic and social database about the Atlantic maritime regions of Ireland, Portugal, the United Kingdom, Spain and France. The database must help constructing an atlas of Atlantic maritime regions online. This project thus consists of evaluating the activities of a maritime basin and in measuring its influence on the onshore and offshore activities that depend on it: for example, what is the volume of maritime freight between Atlantic regions? What is the importance of employment in maritime activities as a function of their proximity to the coast? Marnet will assess the accessibility of relevant data.

4.8. Ecosystems and costs of degradation

The incompleteness of the national accounts, limited to market values indicators, is an already old observation. The question of the economic measure of the well-being of economic agents and the need to incorporate non-merchants elements has emerged in the 1970s. The reflection has expanded to the construction of indicators supplementing the gross domestic product and taking into account the "non-market" value of a protected natural area, the value of the use of time, that of human health and education (see especially [NOR 72, NOR 00, STI 09]).

42 http://marnetproject.eu/.

Concerning the subject of the chapter, this question applies to the value of coastal areas exploited by usages without monetary counterpart (for example coastal leisure): can the cost of management measures and of protection be related to the social benefit of the protection of these areas or with a social cost induced by the loss of a coastal site with essential ecological functions? This type of question generates a large body of economic literature. The satellite accounts mentioned above (see section 4.3 above) just partially answer to them. Their objectives remain limited in terms of non-market values related to the environment[43].

This section briefly recalls the economic methods used to define and estimate the elements of non-market value associated with a natural site and with the environmental "services" that it provides. It focuses on the application of these methods in the case of coastal and marine areas.

4.8.1. *Ecosystem approach to the costs of degradation*

The approach of the value of the environment based on "ecosystem services" is taking an increasing importance in the economic studies supporting the management of the environment. It plays an important role in the discussions about the social and economic assessment required for the MSFD (see [WGE 10]), but the marine realm is an application among others.

The notion of "degradation cost" is defined as the loss for economic agents of the benefits from "ecosystemic services". These are in turn defined as the benefits that humans get from the functioning of ecosystems. These "services" have been the subject of many analyses and discussions in economic journals and expert

43 However, the French transports account brings together sustainable development indicators related to accident analysis, greenhouse gas emissions and other pollutants from transport, as well as the impact of transportation infrastructures.

reports since their first uses at the end of the 1960s[44]. The reports of the *Millennium Ecosystem Assessment* [MIL 05], an expertise group active from 2001 to 2005 under the auspices of the general secretariat of the United Nations, have drawn the inventory of these services, described their developments, the growing risks of alteration or disappearance and proposed measures of protection. These services are classified into four groups:

– provisioning services from resources and habitat (food resources, fresh water, timber, hydrocarbons);

– regulating services (of the climate, floods, weather, water, and health protection);

– cultural services through aesthetic, recreational, cultural, spiritual, educational, and scientific information;

– supporting services, necessary to the three previous groups (natural cycles of nutrients, soil formation, primary production, nitrogen fixation, carbon sequestration).

The benefit approach of ecosystem services, as well as the typology of the MEA, has become a reference that is often used[45]; the "biodiversity strategy 2020" adopted by the EC in 2011 plans to collaborate with member states on such an assessment applied to their territories and "the integration of these values [of ecosystem services] into accounting and reporting systems at EU and national level"[46]. The European works could therefore be oriented toward an operational application of these concepts. The British Department of the Environment and Rural Affairs has already published a guide

44 See, in particular, [TUR 96] for an analysis of ecosystem services focused on coastal areas; [REE 98] for the fundamental problems posed by the methods of evaluation of these services; [TUR 03] and [CHE 09] for critical summaries of the usual methods of ecosystem assessment. Turner *et al.* [TUR 10] offer a summary of the methods of economic and social assessment applied to maritime policy and based on an ecosystem approach.
45 For example, see the report [TEE 10]: expert report under the aegis of the United Nations Environment Program, co-financed by the EC and several European states, and which resumes the categories of services proposed by the MEA.
46 See [COM 11b]: target 2, action 5. A mapping of ecosystems and services is also recommended. See also discussion notes [MAE 13].

[DEF 07] meant for managers for the valuation of ecosystem services, which is based on the most commonly used methods.

In practice, the ecosystem approach of the degradation cost consists, from an analysis of the uses and the functions of ecosystems, of estimating the value assigned to ecosystem services by the end users and non-users, which may be much higher than the actual cost of access and usage of these services. The commercial activities fall within these uses, among others. Starting from the elements of values, the "total economic value" (TEV) of ecosystemic services[47] can be defined, which is usually divided into the following:

– direct use value (benefits obtained from the production of resources and informations: food, recreation, education);

– indirect use value (benefits obtained from the habitat for the resources and the regulating and support services);

– non-use value (value of ecosystems' existence, bequest value, altruistic value[48]).

The cost of degradation corresponds to a loss of profits; it is estimated by the variation of the TEV from a state of the ecosystem into a modified state. In the case of the initial assessment of the MSFD, for example, the evaluation could be done from a voluntarist "return to good status" scenario in 2020 and a "baseline" scenario of evolution of the marine environment with unchanged management.

An important step of the approach consists of valuing the above benefits with the help of indirect methods of preference revelation:

– the contingent valuation method consists of asking a relevant sample of a population about its willingness to pay for the maintenance of a natural site or a resource providing a given ecosystem service or on its willingness to receive compensation in the absence of the possible maintenance of the site;

---

47 For example, see [TUR 10, Chapter 4] and [TUR 96].

48 Existence value: associated, for example, to an outstanding site just because it exists. Bequest value: associated with the interest in transmitting a preserved ecosystem to future generations. Altruistic value: associated with the interest in seeing the ecosystem produce services for other users.

– the travel cost method consists of evaluating the time and travel cost expenses for end users who visit a natural site (beach, leisure fishing area, etc.) and the extra cost consented if, after degradation of the site, they had to find a substitution site further away;

– the hedonic pricing method consists of estimating the economic value of a natural site or an ecological function from the market price of an asset whose value is supposed to depend on the state of the site; for example the price of land in the vicinity of the site.

This set of values is used to assess, as an alternative (from an initial state of the ecosystem to a altered state), the degradation costs and the benefits of restoring ecosystem services through the variation in value of the benefits, which derive from them (for example the increase in fishery resources after the closure of geographically defined areas, or the variation of microbiological concentrations in marine waters after local earthmoving and disappearance of a wetland area). The latter are compared to the economic impacts of management choices or replacement activities, including a compensation for degraded ecosystems. The technical costs of restoration projects are also to be taken into account. This set of losses and gains is integrated in a cost-benefit analysis[49].

The ecosystem approach, if it is systematized, will lead to growing needs in database and metadata management that are still difficult to assess. The starting point of this tendency: the database *Environment Valuation Reference Inventory*, cataloging of empirical studies on the economic value of environmental benefits and of the effects on human health, has been put in place toward the end of the 1990s by the Canadian Environment Department[50]. It is the subject of an international collaboration in which homologous American, Australian, British, French, New Zealand government departments participated.

49 On the basis of available data, cost-benefit analysis can be replaced by other methods. For example, cost-effectiveness analysis consists of finding the option of least cost or higher benefit that allow achieving specific management objectives by maintaining, compensating or restoring all or part of the ecosystem functions.
50 www.evri.ca.

4.8.2. *Practical application of the ecosystem approach*

In this section and in the following section, the approach to ecosystem degradation costs is tested on two practical applications. The Pendleton report [NEP 09] is discussed here.

Published by a group of American scholars, this handbook provides coastal site restoration managers with the economic information necessary for the design and the monitoring of projects in this domain. Focused on the evaluation of potential economic benefits from the restoration of coastlines, it analyzes, on the one hand, the market value generated by coastal and marine activities in the United States, and, on the other hand, the benefits derived from restoring estuarine areas, which constitute the vast majority of U.S. coastal areas. It includes:

– an analysis of the employment and value added of coastal activities based on NOEP data (see above);

– an analysis of the value of ecosystem services, consistent with the approach of [MIL 05], to coastal and beach recreation areas and water sports;

– a benefit analysis, for determined categories of users, resulting from the protection or restoration of certain resources or parts of the coastline. For example, fisheries resources for commercial fishing; coastal areas exposed to erosion, flooding, tidal waves and hurricanes; seaport areas; protection of estuaries; and effects on the price of land.

The main information used is of three types:

– estimates of accounting indicators of the local economy (NOEP does not collect, but estimates local data);

– estimates of the number of visits for beach recreation and water sports;

– estimates of users' willingness to pay for the protection of estuarine and coastal areas (as per survey data).

The results of sector-based analysis specify the value added and employment by activity, state and county. With respect to the other two types of information, the data and the results produced by various studies of sites allow us to estimate the benefits and willingness to pay in the total population of end users. In unfavorable cases, they may vary in very wide intervals. In the case of some types of recreation where data are missing, results are simply qualitative. This work brings forward the interest in continuing with studies of sites.

4.8.3. *Direct approach to degradation costs*

In the inventory of usable methods for the initial assessment of the MSFD, WG-ESA [WGE 10] mentioned the approach advocated by the participating French team, which was based on a direct analysis of degradation costs. A practical argument in favor of this method was its speed of implementation, given the limited time allocated to the initial assessment. The approach uses a set of costs of which several are observable:

– monitoring and information costs (e.g. coastal water monitoring systems, observatories, oceanography studies and research, regulation);

– curative and preventive costs, i.e. for marine environment preservation and treatment of degradation (e.g. biodiversity protection, coastline restoration, accelerated phasing-in of double-hull oil tankers);

– mitigation costs of the impacts on populations: curative costs incurred after degradation (for example collection of green algae after harmful algal bloom);

– "residual" costs due to the remaining impacts, after mitigation, of the degradation (loss of direct benefits from marine resources produced by ecosystems, for example a decrease in tourist businesses' or commercial fisheries' incomes).

These four components provide a comprehensive picture of the costs of degradation, usable in a cost-benefit analysis, but quantifiable under the condition that data are available.

The approach has been implemented in the framework of the MSFD initial assessment for France: fact sheets dealing with the costs associated with several types of pressures (including: marine waste, pathogenic microbes, eutrophication, biodiversity, fishery resources, illegal discharges of oil by ships at sea) are attached to the above-mentioned APME. They provide estimates for each of the four types of above-mentioned costs (see Table 4.5) and thus allow us to assess the amount of data required for this type of assessment.

Actions	Costs (thousands of euros)
(1) Monitoring and collection of information including:	26,604
- administrative services and public agencies in charge of marine environment protection	6,437
- impact studies on aggregate extraction and harbor works	772
- professional observatories	2,915
- volunteering observatories	537
- local NGOs	768
- research	15,175
(2) Positive actions including:	11,600
- administrative services and public agencies in charge of marine environment protection	6,739
- NGOs, mainly national	54
- protected areas	4,807
(3) Mitigation including:	15,845
- administrative services and public agencies in charge of marine environment protection	5,667
protected areas	782
- mitigation and compensation measures for aggregate extraction and harbor works	9,396

Table 4.5. *Cost of biodiversity degradation. English Channel–North Sea subregion (source: [MED 12b]: Initial assessment of marine waters. Economic and social analysis, sheet 9: Levrel H., Jacob C. Costs related to the loss of biodiversity and seabed environmental quality)*

The results of the "biodiversity" sheet presented in Table 4.5 cover the different categories of degradation costs, but several other sheets present partial evaluations; missing data have made the quantification of some categories impossible.

4.8.4. *Combining the different approaches: the problem of local data*

The MSFD initial economic assessment and the Pendleton report's analysis of the value of coasts and estuaries allow testing on operational cases three approaches to the economic value of ecosystems and their uses: the accounting value produced by maritime activities; the benefits derived from ecosystem services; the direct analysis of the costs of degradation in the case of France. The question is about how the results of these three approaches could be combined into a global evaluation.

On the methodological level, production and maritime employment by activity, derived from national accounts, have the advantage of being observable indicators, measuring the economic importance of maritime activities "specific" and "linked" to the maritime economy, as defined above (section 4.4). If the problem shifts from maritime activities in general to market and non-market benefits from coastal resources and natural sites, the economic literature will move its focus to the concept of ecosystem services; the latter includes, among other things, certain components of maritime production, in particular beach and water recreation and sports, which may provide benefits to users higher than the payments they make for that use. The valuation is then performed by other estimation techniques than the direct observation of market prices. By definition, the TEV also incorporates the costs of degradation directly estimated. From this standpoint, the three approaches can be combined, one encompassing the other two.

However, the differences between these methods of evaluation are numerous at their implementation level and three of them are important.

First, the accounting analysis of production and employment is based on a classification of activities that is partially adapted to maritime activities. The direct approach of the degradation costs is based on a set of management costs that are not defined to achieve an exact partition of the overall costs and do not add up. The difficulty is then due to the fact that the corresponding management measures may have very broad objectives in terms of environment quality, not especially aiming at a specific case of degradation: the costs are likely to be impossible to relate to this specific case. The ecosystem services approach, for its part, is facing difficulties frequently discussed in the economic literature about the classification of benefits derived from services[51]. It requires an in-depth knowledge of the effects of anthropic pressures on the ecosystem functions in the studied cases not only because of complex cumulative phenomena of some contaminants, nonlinear responses of ecosystems to pressures and viability threshold effects of these ecosystems, but also because of the numerous and simultaneous pressures on the environment of a given area, of which each of the specific effects, ideally, would have to be identified.

Second, local data requirements are a problem for all three approaches: the initial assessment of the MSFD has brought forward the critical need for local data, which conditions an effective implementation of the analysis of maritime activities. Ecosystem services, as it has been seen, depend on their local conditions of operation; the valuation of services must therefore be specified by sites or associated with spatial references (see [DEF 07, Chapter 3] and [NEP 09, Chapter 8]). The costs of degradation depend on uses and the types of pressure; they also have a spatial dimension. Any cost-benefit analysis of uses and ecosystems services at the sea-basin scale thus requires a significant volume of local data which, at this stage, is not available.

Local data, when they are available, also raise the difficulty of their level of resolution: in the case of France, the observable local data of

51 For clarifications on this point, see [TUR 10, Chapter 4] and [CHE 09, Chapter 5].

maritime activities exist and relate to municipalities when they are provided by INSEE (see section 4.5). These administrative units certainly differ from the spatial units of homogeneous marine ecosystems. The most realistic option is to accept this discrepancy and to combine local data at different levels of resolution.

Third, the costs and benefits resulting from the three approaches may be attributable to different categories of agents. For example, an industrial activity may yield goods and create jobs by degrading a coastal site to the detriment of benefits to other agents dependent on the quality of the impacted environment (residual costs); therefore, at the expense of other types of jobs associated with their activities. Restoration and monitoring are carried out by public agencies (curative costs and costs of monitoring), creating jobs in other categories. From the operational perspective, it may be assumed that these differences in categories of agents have an economic, social and political significance for managers and that they are relevant to study. From the methodological perspective, they could only be correctly analyzed using local data with a national coverage.

Hence, a correct combination of the three approaches is faced with two difficulties:

– the difficulty in establishing a one-to-one correspondence between water uses, marine ecosystem services and costs of degradation, using a set of quantitative indicators;

– the lack of local data that would allow to value the benefits and losses of the different categories of agents per specific spatial area, even in accepting significant errors or deviations.

That is why an adequate combination of the approaches should not aggregate the available data in an undifferentiated total, but maintain the traceability of benefits and losses per category of agents and per spatial area. The downside of this option is the large amount of data required by a comprehensive assessment which, given the set of available information, remains out of reach. This work has begun, in more realistic proportions, in the studies referred to in the previous sections. It has also begun in a case, which has not been detailed here, namely the studies of impacts of oil spills (for example [GRI 86]), carried out with the objective of compensating the impacted economic activities and specific riparian groups or assessing non-market amenity losses affecting certain categories of end users.

4.9. Conclusions

On the basis of dossiers related to the current phase of the action of the EC, the present contribution has focused on a general and enduring problem and highlights the main aspects of the European maritime policy and its need for economic evaluation.

This policy, devoid of specific legal basis in the treaty of the EU, concerns a high diversity of economic activities. It has been built by successive layers relating to a range of economic activities and environmental measures. The needs for data that each policy initiative puts systematically forward depend on competitiveness and growth objectives for maritime industries as well as objectives of sustainability and mitigation of the ecological footprint. Under optimal conditions, these data would allow comparing the value added and the number of jobs generated by maritime activities with the estimated benefits generated by the protection of marine ecosystems and, conversely, the losses due to their degradation. The question is then to what extent the data available allow study in this direction.

The methods of maritime economic data collection and processing described in this contribution allow confronting this question on the basis of practical case studies. These methods are used in several countries, but also by the EC, which seeks to combine the two objectives of "blue growth" and return to the "good status" of marine waters and environment. These two objectives require the analysis of two different channels of value production: economic activities and maritime employment, on the one hand, and marine ecosystem services, on the other hand.

The analysis has been attempted in this chapter, using studies on economic indicators applied to maritime activities and basins, the benefits from ecosystem services and the costs of degradation of marine ecosystems. Several of these works, in Europe and the United States, have provided methods actually applied to procedures of public decision making (among other things, the MSFD). They also highlight two implementation difficulties, closely related to one another. First, the difficulty to accurately describe maritime activities, marine ecosystem services and management costs by a set of indicators: if their economic and ecological nature is feasible to approach, they remain difficult to quantify by simple methods. Second, the lack of local data that enables quantifying these activities, services and costs at a sufficiently fine resolution level. This lack of data is related to the coherence of economic indicators as well as to collection costs; it is also related to the complexity of ecosystem functions whose associated benefits are to be valued. Therefore, it is a complex task to ensure a coherent spatial coverage of these economic activities, benefits and costs at the scale of a coastal site, a seafront or a maritime basin, unless at the price of data gaps that are to be progressively filled.

One of the ways to achieve this is to continue with local site studies. When they are sufficient in number, they make it possible to estimate the above economic values and improve knowledge on local ecosystem services and uses. A proper collection of economic indicators at the scale of a seafront or a basin also requires geographic information data on uses and ecosystems. Improving knowledge on maritime basins and the main economic assessment methods requires continuing with, therefore, data collection in the long term; in the short term, it is only possible to compromise within the limits of available information.

4.10. Appendices

	USA	Italy	GB	France	Canada	Schleswig-Holstein	Australia	New Zealand	Netherlands	Ireland	Spain
Sources	Noep 2009	Censis 2011	Pugh 2007	Ifremer 2011	Gardner Pinfold 2009	MC 2005	Allen CG2005	Statistics NZ 2006	Wijnolst 2003	Univ. Galway 2010	Innovamar 2011
Specific											
Fishing	x	x	x	x	x	x		x	x	x	x
Aquaculture	x	x	x	x	x	x		x		x	x
Sea salt											x
Marine aggregates	x		x	x							
Offshore oil and gas	x		x		x			x		x	x
Coastal power plants (1)				x							x
Marine energy			x			x				x	
Maritime work, inc. offshore platforms			x	x	x	x			x		
Submarine cables			x	x							
Submarine oil pipelines, gas pipelines											
Maritime transport, inc. cruises	x	x	x	x	x	x		x	x	x	x
Ports management and services to vessels (2)	x	x	x	x	x	x		x	x	x	
Marinas	x	x	x	x	x	x	x		x		x

Recreational fishing					x	x					
National navy, defense		x	x	x	x	x		x	x		
Marine science, research			x	x	x	x		x	x		x
Operational oceanography and marine data			x	x	x	x		x		x	
Route survey											
Coastal water monitoring				x	x						
	USA	Italy	GB	France	Canada	Schleswig-Holstein	Australia	New Zealand	Netherlands	Ireland	Spain
Linked – essential											
Seafood trade	x	x		x		x		x		x	
Seafood processing	x	x	x	x	x	x		x		x	x
Ship and boat building	x	x	x	x	x	x	x	x	x	x	x
Ship and boat repair and maintenance	x	x	x	x	x		x	x	x	x	x
Ship dismantling									x		x
Marine equipment manufacturing (3)	x		x			x	x		x	x	x
Boat equipment (distribution)	x		x			x	x	x		x	x
Offshore oil and gas services				x	x	x		x	x		
Port services (4)		x	x	x	x	x		x	x	x	x

Port handling	x	x	x	x	x			x		x	x
Warehousing	x	x	x			x					
Rental and leasing of navigation equipment.	x	x	x	x				x	x	x	
Offshore supply services									x		
Brokerage				x			x		x	x	
Classification, control			x			x			x		
Marine insurance		x	x	x		x			x	x	
Banks			x			x				x	
Accommodation	x		x	x	x						
Restaurants and beverage serving activities	x	x	x	x	x						
Boating-related services	x	x	x	x	x	x	x		x	x	x
Tourist-related services (other than boating services)	x	x	x	x	x						x
Signaling, safety, rescue		x	x	x	x			x	x	x	
Food safety control					x						
Education, training			x	x		x		x	x	x	x
Coastal and marine environment protection			x	x	x			x			

	USA	Italy	GB	France	Canada	Schleswig-Holstein	Australia	New Zealand	Netherlands	Ireland	Spain
Linked – complementary											
Biotechnology										x	
Professional textiles and clothing		x									
Misc. naval equipment											
River works				x		x					
Inland navigation				x		x			x		x
Travel agencies		x		x	x						x
Other tourist services		x		x	x						
Wastewater and waste treatment											x
Shipping press						x					
Legal maritime services						x				x	
Industry associations and other similar maritime bodies						x					

(1) Employment only.
(2) Port authority, pilotage, berthing and towing.
(3) Of which fishing gear.
(4) Services to vessels, goods and passengers, including freight forwarders, shipping agents and customs. For France, employment data only; no accounting data available.

Table 4.6. *Maritime activities taken into account in national maritime economy reports*

Code	Detail
A	Agriculture, forestry and fishing
B	Extraction and mining
C	Manufacturing
D	Electricity, gas, steam and air conditioning supply
E	Water supply, sewerage, waste management and remediation activities
F	Construction
G	Wholesale and retail trade, repair of motor vehicles and motorcycles
H	Transportation and storage
I	Accommodation and food service activities
J	Information and communication
K	Financial and insurance activities
L	Real estate activities
M	Professional, scientific and technical activities
N	Administrative and support service activities
O	Public administration
P	Education
Q	Human health and social work activities
R	Arts, entertainment and recreation
S	Other service activities
T	Activities of households and employers; undifferentiated goods- and services-producing activities of households for own use
U	Activities of extraterritorial organizations and bodies

Table 4.7. *French nomenclature of activities, revision 2008 Level 1: list of sections (Source: INSEE)*

4.11. Bibliography

NOTE.– French and European legislative texts mentioned in the footnotes are not included in the references below.

[ADE 12] ADEC, Société de développement économique du Saint-Laurent (Sodes), Transports Québec, *Etude de l'impact économique de l'industrie maritime au Québec*, SODES, Québec, 2012.

[ALL 05] ALLEN CONSULTING GROUP PTY LTD, *Benchmarking Australia's Marine Industries*, Tourism and Resources, Department of Industry, Canberra, 2005.

[BON 02] BONNIEUX F., RAINELLI P., "Evaluation des dommages des marées noires: une illustration à partir du cas de l'Erika et des pertes d'agrément des residents", *Economie et statistique*, vol. 357–358, pp. 173–187, 2002.

[BOO 12] BOOZ ALLEN HAMILTON, NOAA coastal services center, ENOW Final Economic Report, NOAA CSC, Charleston, 2012.

[CEN 11] CENSIS, IV Rapporto sull'economia del mare. Cluster marittimo e sviluppo in Italia e nelle regioni, Federazione del Mare, Sintesi, Roma, 2011.

[CHE 09] CHEVASSUS-AU-LOUIS B., SALLES J.M., PUJOL J.L., Approche économique de la biodiversité et des services liés aux écosystèmes, Centre d'analyse stratégique, Paris, 2009.

[COL 13] COLLET I., "Economic ebb and flow in maritime sectors", *Statistics in Focus*, no. 5, Eurostat, Luxembourg, 2013.

[CON 13] CONSEIL ÉCONOMIQUE, SOCIAL ET ENVIRONNEMENTAL, *L'Extension du plateau continental au-delà des 200 milles marins: un atout pour la France, Avis du CESE sur le rapport présenté par M.G. Grignon*, Les Editions des Journaux Officiels, Paris, 2013.

[DEF 07] DEFRA DEPARTMENT FOR ENVIRONMENT, FOOD AND RURAL AFFAIRS, *An Introductory Guide to Valuing Ecosystem Services*, Defra Publications, London, 2007.

[ECO 07] ECORYS NEDERLAND BV, Ex ante evaluation of the TEN-T multi-annual program 2007–2013, Final report-2, Ecorys, Rotterdam, Brussels, October 2007.

[ECO 12] ECORYS NEDERLAND BV, Blue growth. Scenarios and drivers for sustainable growth from oceans, seas and coasts, MARE/2010/01, Ecorys, Rotterdam, Brussels, 2012.

[EFT 12] EFTEC, Synthesis of the initial observations on economic and social analysis in draft MSFD initial assessment reports in the OSPAR region, EFTEC, London, 2012.

[EUR 01] EUROPEAN COMMISSION, White Paper. European transport policy for 2010: time to decide, COM(2001) 370 final, Brussels: EC, 12 September 2001.

[EUR 02] EUROPEAN COMMISSION, Communication from the commission to the European parliament, the council, the economic and social committee and the committee of the regions. Towards a global partnership for sustainable development, COM(2002) 82 final, Brussels: EC, 13 February 2002.

[EUR 03] EUROPEAN COMMISSION, Communication from the commission to the European parliament and the council. The commission's legislative and work programme for 2004, COM(2003) 645 final, Brussels: EC, 29 October 2003.

[EUR 07a] EUROPEAN COMMISSION, Green paper on better ship dismantling (presented by the commission) , COM(2007) 269 final, Brussels: EC, 22 May 2007.

[EUR 07b] EUROPEAN COMMISSION, Communication from the commission to the European parliament, the council, the EESC and the CR. An integrated maritime policy for the EU, COM(2007) 575 final, Brussels: EC, 10 October, 2007.

[EUR 07c] EUROPEAN COMMISSION, Invitation to tender for the supply of statistical services. Study in the field of maritime policy, Tender specifications 2007/S 179-218229, OJ/S S179, Luxembourg, 18 September 2007.

[EUR 08] EUROPEAN COMMISSION, Communication from the commission to the European parliament, the council, the economic and social committee and the committee of the regions. An EU strategy for better ship dismantling COM(2008) 767 final, Brussels: EC, 19 November 2008.

[EUR 09a] EUROPEAN COMMISSION, Communication from the commission to the European parliament, the council, the economic and social committee and the committee of the regions. Strategic goals and recommendations for the EU's maritime transport policy until 2018, COM(2009) 8 final, Brussels: EC, 21 January 2009.

[EUR 09b] EUROSTAT, "Nearly half of the population of the EU countries with a sea border is located in coastal regions", *Statistics in Focus*, no. 47, Eurostat, Luxembourg, 2009.

[EUR 10] EUROPEAN COMMISSION, Communication from the commission. Europe 2020. A strategy for smart, sustainable and inclusive growth, COM(2010) 220 final, Brussels: EC, 3 March 2010.

[EUR 11a] EUROPEAN COMMISSION, White paper. Roadmap to a single European transport area. Towards a competitive and resource-efficient transport system, COM(2011) 144 final, Brussels: EC, 28 March 2011.

[EUR 11b] EUROPEAN COMMISSION, Communication from the commission to the European parliament, the council, the economic and social committee and the committee of the regions. Our life insurance, our natural capital: an EU biodiversity strategy to 2020, COM(2011) 244 final, Brussels: EC, 3 May 2011.

[EUR 11c] EUROPEAN COMMISSION, Communication from the commission to the European parliament, the council, the economic and social committee and the committee of the regions. Reform of the common fisheries policy, COM(2011) 417 final, Brussels: EC, 13 July 2011.

[EUR 11d] EUROPEAN COMMISSION, Proposal for a regulation of the European parliament and of the council on union guidelines for the development of the trans-European transport network, COM(2011) 650 final/2, Brussels: EC, 19 December 2011.

[EUR 12] EUROPEAN COMMISSION, Communication from the commission to the European parliament, the council, the economic and social committee and the committee of the regions. Blue growth. Opportunities for marine and maritime sustainable growth, COM(2012) 494 final, Brussels: EC, 13 September 2012.

[EUR 13a] EUROPEAN COURT OF AUDITORS, Have the Marco Polo programmes been effective in shifting traffic off the road?, Special Report No. 3, Publication Office of the European Union, Luxembourg, 2013.

[EUR 13b] EUROSTAT, Agriculture, forestry and fishery statistics, KS-FK-13-001-EN-C, Publication Office of the EU, Luxembourg, 2013.

[FRA 06a] FRAUNHOFER INSTITUTE SYSTEMS AND INNOVATION RESEARCH, Compete analysis of the contribution of transport policies to the competitiveness of the EU economy and comparison with the United States, Final report, version 2.1, Fraunhofer Institute, Karlsruhe, October 2006.

[FRA 06b] FRAUNHOFER INSTITUTE SYSTEMS AND INNOVATION RESEARCH, Compete Annex 7. Development of productivity in the transport sector, version 2.0, Fraunhofer Institute, Karlsruhe, 18 August 2006.

[GAR 09a] GARDNER PINFOLD CONSULTING ECONOMISTS LTD, *Economic Impact of Marine Related Activities in Canada*, Fisheries and Oceans Canada, Ottawa, March 2009.

[GAR 09b] GARDNER PINFOLD CONSULTING ECONOMISTS LTD, *Economic Impact of the Nova Scotia Ocean Sector 2002–2006*, Gardner Pinfold CEL, Halifax, 2009.

[GRI 86] GRIGALUNAS T.A., ANDERSON R.C., BROWN G.M., *et al.*, "Estimating the cost of oil spills: lessons from the Amoco Cadiz incident", *Marine Resource Economics*, vol. 2, no. 3, pp. 239–262, 1986.

[IFR 09] IFREMER, BALANCE TC, *Study in the Field of Maritime Policy. Approach towards an Integrated Maritime Policy Database*, vol. 1–2, EC, Brussels, 2009.

[INN 11] INNOVAMAR, *Cuantificación Económica del sector marítimo y su desagregación sectorial*, Innovamar, Madrid, 2011.

[JAR 08] JAROWINSKY M., *Fortschreibung der Potenzialanalyse für die maritime Wirtschaft in Schleswig-Holstein*, MC Marketing Consulting, Kiel, 2008.

[KAL 10] KALAYDJIAN R. (ed.), *French Marine Economic Data 2009*, Ifremer, Brest, France, 2010.

[KIL 00] KILDOW J.T., BAIRD B., COLGAN C.S., *et al.*, "Improving economic information for ocean and coastal resources", *Proceedings of the 17th International Conference of the Coastal Society "Coasts at the Millennium"*, Portland, pp. 305–313, 2000.

[KIL 09] KILDOW J.T., COLGAN C.S., SCORSE J., *State of the U.S. Ocean and Coastal Economies*, NOEP, Monterey, CA, 2009.

[LUG 91] LUGER M.I., "The economic value of the coastal zone", *Journal of Environmental Systems*, vol. 21, no. 4, pp. 279–301, 1991.

[MAE 13] MAES J., TELLER A., ERHARD M., *et al.*, Mapping and assessment of ecosystems and their services. An analytical framework for ecosystem assessments under action 5 of the EU biodiversity strategy to 2020, Technical Report 2013-067, Publication office of the European Union, Luxembourg, 2013.

[MIL 05] MILLENNIUM ECOSYSTEM ASSESSMENT (MEA), *Ecosystems and Human Well-Being: Synthesis*, Island Press, Washington, 2005.

[MED 12a] MEDDE, "Préfecture maritime Méditerranée, préfecture de région Provence-Alpes-Côte d'Azur", *Plan d'action pour le milieu marin sous-région marine Méditerranée occidentale. Evaluation initiale des eaux marines*, DIRM Méditerranée, Marseille, 2012.

[MED 12b] MEDDE, "Préfecture maritime Manche et mer du Nord, préfecture de région Haute-Normandie", *Plan d'action pour le milieu marin sous-région marine Manche-mer du Nord. Evaluation initiale des eaux marines*, DIRM Manche Est-mer du Nord, Le Havre, 2012.

[MON 14a] MONACO A., PROUZET P. (eds.), *Ecosystem Sustainability and Global Change*, ISTE, London, and John Wiley & Sons, New York, 2014.

[MON 14b] MONACO A., PROUZET P. (eds.), *Development of Marine Resources*, ISTE, London, John Wiley & Sons, New York, 2014.

[MOR 10] MORRISSEY K., HYNES S., CUDDY M., *et al.*, *Ireland's Ocean Economy*, National University of Ireland, Galway, 2010.

[NOR 72] NORDHAUS W.D., TOBIN J., "Is growth obsolete?", *Economic Research: Retrospect and Prospect*, vol. 5, no. 96, pp. 1–80, 1972.

[NOR 00] NORDHAUS W.D., "New directions in national economic accounting", *American Economic Review*, vol. 90, no. 2, pp. 259–263, May 2000.

[OXF 13] OXFORD ECONOMICS, "The economic impacts of the UK maritime services sector", Report for Maritime UK, Oxford Economics, Oxford, 2013.

[PEN 09] PENDLETON L.H., *The Economic and Market Value of America's Coasts and Estuaries: What's at Stake?*, Coastal Ocean Values Press, Washington, 2009.

[PON 80] PONTECORVO G., WILKINSON M., ANDERSON R., *et al.*, "Contribution of the ocean sector to the United States economy", *Science*, vol. 208, no. 4447, pp. 1000–1006, 1980.

[POL 00] POLICY RESEARCH CORPORATION NV (PRC), INSTITUTE OF SHIPPING ECONOMICS AND LOGISTICS (ISL), Economic Impact of Maritime Industries in Europe, European Commission, Brussels, 2000.

[PUG 08] PUGH D., *Socio-Economic Indicators of Marine-Related Activities in the UK Economy*, The Crown Estate, London, 2008.

[REE 98] REES W.E., "How should a parasite value its host?", *Ecological Economics*, vol. 25, pp. 49–52, 1998.

[SEC 07] SECRETARY-GENERAL OF THE EUROPEAN COMMISSION, Commission staff working document. Accompanying document to the communication from the commission to the EP, the council, the EESC and the CR. An integrated maritime policy for the EU, SEC(2007) 1278, Brussels, 10 October 2007.

[SEC 10] SECRETARY-GENERAL OF THE EUROPEAN COMMISSION, Commission staff working document. Lisbon strategy evaluation document, SEC(2010) 114 final, Brussels, 2 February 2010.

[SEC 11] SECRETARY-GENERAL OF THE EUROPEAN COMMISSION, Commission staff working document. Impact assessment accompanying the document: proposal for a regulation of the EP and of the Council on Union Guidelines for the development of the trans-European transport network, SEC(2011) 1212 final/2, Brussels, 19 December 2011.

[STA 06] STATISTICS NEW ZEALAND, "New Zealand's marine economy 1997–2002", *Environmental Series, Statistics* Wellington, New Zealand, 2006.

[STI 09] STIGLITZ J.E., SEN A., FITOUSSI J.P., Report of the commission on the measurement of economic performance and social progress, INSEE, CMEPSP, Paris, 2009.

[TEE 10] TEEB: THE ECONOMICS OF ECOSYSTEMS AND BIODIVERSITY, The economics of ecosystems and biodiversity: mainstreaming the economics of nature: a synthesis of the approach, conclusions and recommendations of TEEB, United Nations Environment Program, Nairobi, 2010.

[TUR 96] TURNER R.K., SUBAK S., ADGER W.N., "Pressures, trends and impacts in coastal zones: interactions between socioeconomic and natural systems", *Environmental Management*, vol. 20, no. 2, pp. 159–173, 1996.

[TUR 03] TURNER R.K., PAAVOLA J., COOPER P., *et al.*, "Valuing nature: lessons learned and future research directions", *Ecological Economics*, vol. 46, pp. 493–510, 2003.

[TUR 10] TURNER R.K., HADLEY D., LUISETTI T., *et al.*, *An Introduction to Socio-Economic Assessment within a Marine Strategy Framework*, Department for Environment, Food and Rural Affairs, London, 2010.

[UN 08] UN STATISTICS DIVISION, *EC, OECD, World Tourism Organization, 2008 Tourism Satellite Account: Recommended Methodological Framework,* UN, New York, 2008.

[WIJ 03] WIJNOLST N., JENSSEN J.I., SØDAL S., *European Maritime Clusters. Global Trends, Theoretical Framework, the Cases of Norway and the Netherlands, Policy Recommendations*, Delft University Press, Delft, 2003.

[WGE 10] WG-ESA, Working group on economic and social assessment, Economic and Social Analysis for the Initial Assessment for the Marine Strategy Framework Directive: A Guidance Document, European Commission, Brussels, 21 December 2010.

List of Authors

Pedro CABRAL
ISEGI/UNI
Lisbon
Portugal

Christian CHABOUD
IRD – UMR EME
Sète
France

Pierre FRÉON
IRD – UMR EME
Sète
France

Régis KALAYDJIAN
Ifremer – UEM
Issy-les-Moulineaux
France

Harold LEVREL
AMURE
Ifremer
Plouzané
France

Océane MARCONE
AMURE
Ifremer
Plouzané
France

André MARIOTTI
University Pierre and Marie Curie
Paris
France

André MONACO
Cefrem/CNRS-UPVD
Perpignan
France

Rémi MONGRUEL
AMURE
Ifremer
Plouzané
France

Jean-Charles POMEROL
University Pierre and Marie Curie
INSIS/CNRS
Paris
France

Patrick PROUZET
Ifremer
Issy-les-Moulineaux
France

Sylvestre VOISIN
Arcansys
Paris
France

Index

A, B, C

D, E

F, G

I, L

M

O, P, Q, R

S, T, W

Other titles from

iSTE

in

Oceanography and Marine Biology

2014

Seas and Oceans Set coordinated by MONACO André, PROUZET Patrick

Ocean in the Earth System

The Land-Sea Interactions

MUTTIN Frédéric

Marine Coastal and Water Pollutions: Oil Spill Studies

2013

LÉCUYER Christophe

Water on Earth: Physicochemical and Biological Properties

Printed and bound by CPI Group (UK) Ltd, Croydon, CR0 4YY
13/09/2026
14961077-0004